The Evolution of Agency and Other Essays

This book presents a collection of linked essays on the topic of biological evolution written by one of the leading philosophers of biology, Kim Sterelny.

The first half of the book explores most of the main theoretical controversies about evolution and selection. Sterelny argues that genes are not the only replicators: non-genetic inheritance is also extremely important, and is no mere epiphenomenon of gene selection. The second half of the book applies some of these ideas in considering cognitive evolution. Concentrating on the mental capacities of simpler animals rather than those of humans, Sterelny argues for a general distinction between detection and representation, and that the evolution of belief, like that of representation, can be decoupled from the evolution of preference.

These essays, some never before published, form a coherent whole that defends not just an overall conception of evolution but also a distinctive take on cognitive evolution. The volume should be of particular interest to graduate students and professionals of biology, cognitive science, and the philosophy of biology.

Kim Sterelny is Professor of Philosophy at Victoria University of Wellington and Senior Research Fellow in the Philosophy Program at Australian National University.

T0296479

CAMBRIDGE STUDIES IN PHILOSOPHY AND BIOLOGY

General Editor
Michael Ruse   *University of Guelph*

Advisory Board
Michael Donoghue   *Harvard University*
Jean Gayon   *University of Paris*
Jonathan Hodge   *University of Leeds*
Jane Maienschein   *Arizona State University*
Jesús Mosterín   *Instituto de Filosofía (Spanish Research Council)*
Elliott Sober   *University of Wisconsin*

Alfred I. Tauber: *The Immune Self: Theory or Metaphor?*
Elliott Sober: *From a Biological Point of View*
Robert Brandon: *Concepts and Methods in Evolutionary Biology*
Peter Godfrey-Smith: *Complexity and the Function of Mind in Nature*
William A. Rottschaefer: *The Biology and Psychology of Moral Agency*
Sahotra Sarkar: *Genetics and Reductionism*
Jean Gayon: *Darwinism's Struggle for Survival*
Jane Maienschein and Michael Ruse (eds.): *Biology and the Foundation of Ethics*
Jack Wilson: *Biological Individuality*
Richard Creath and Jane Maienschein (eds.): *Biology and Epistemology*
Alexander Rosenberg: *Darwinism in Philosophy, Social Science and Policy*
Peter Beurton, Raphael Falk, and Hans-Jörg Rheinberger (eds.): *The Concept of the Gene in Development and Evolution*
David Hull: *Science and Selection*
James G. Lennox, *Aristotle's Philosophy of Biology*

The Evolution of Agency and Other Essays

KIM STERELNY
Victoria University of
Wellington and
The Australian
National University

CAMBRIDGE
UNIVERSITY PRESS

CAMBRIDGE UNIVERSITY PRESS
Cambridge, New York, Melbourne, Madrid, Cape Town, Singapore,
São Paulo, Delhi, Dubai, Tokyo, Mexico City

Cambridge University Press
The Edinburgh Building, Cambridge CB2 8RU, UK

Published in the United States of America by Cambridge University Press, New York

www.cambridge.org
Information on this title: www.cambridge.org/9780521645379

First published 2001

A catalogue record for this publication is available from the British Library

Library of Congress Cataloging in Publication Data

Sterelny, Kim.   S]
The evolution of agency and other essays / Kim Sterelny.
p. cm. – (Cambridge studies in philosophy and biology)
Includes bibliographical references.
1. Evolution (Biology) – Philosophy. 2. Animal intelligence.
ISBN 0-521-64231-0 – ISBN 0-521-46537-9 (pbk.)
I. Title. II. Series.
QH360.5 .S76 2000 576.8¢01–dc21   00-025317

ISBN 978-0-521-64231-6 Hardback
ISBN 978-0-521-64537-9 Paperback

To Lucy

Contents

ix

Contents

Sources

Chapter 1: Evolution and Agency: A User's Guide. Not previously published.

Chapter 2: The Return of the Gene (with Philip Kitcher). *Journal of Philosophy* 85: 339–60, 1988.

Chapter 3: The Extended Replicator (with Kelly Smith and Mike Dickison). *Biology and Philosophy* 11: 377–403, 1996.

Chapter 4: The Return of the Group. *Philosophy of Science* 63: 562–84, 1996.

Chapter 5: Punctuated Equilibrium and Macro-evolution. *Trees of Life: Essays in Philosophy of Biology.* P. Griffiths. Dordrecht, Kluwer: 41–64, 1992.

Chapter 6: Explanatory Pluralism in Evolutionary Biology. *Biology and Philosophy* 11: 193–214, 1996.

Chapter 7: Darwin's Tangled Bank. Not previously published.

Chapter 8: Where Does Thinking Come From? *Biology and Philosophy* 12: 551–66, 1997.

Chapter 9: Basic Minds. *Philosophical Perspectives* 9: 251–70, 1995.

Chapter 10: Intentional Agency and the Metarepresentation Hypothesis. *Mind and Language* 13: 11–28, 1998.

Chapter 11: Situated Agency and the Descent of Desire. *Biology Meets Psychology: Constraints, Conjectures, Connections.* V. Hardcastle. Cambridge, MIT Press, 1999.

Chapter 12: The Evolution of Agency. Not previously published.

Preface

This work is a collection of twelve essays, nine of which have previously seen the light of print. I fondly imagine that each essay is independently intelligible, but jointly they constitute, I hope, an integrated conception of evolution in general and the evolution of mind in particular. However, that conception is incomplete in important ways; this is research in progress. In particular, like many in the philosophy of biology trade, I have only begun to sketch out the connections between evolutionary theory and other crucial areas of biological thinking; in particular, ecology and developmental biology. The conception of cognitive evolution, too, is still tentative in important respects.

I outline the contents of this collection, pointing to themes, interconnections and revisions in the first chapter. So I will not recapitulate that survey here. Instead, I will use this opportunity to acknowledge the debts I owe others, intellectually, institutionally, and personally. Perhaps first of all I should thank my co-authors, Philip Kitcher, Kelly Smith and Mike Dickison, for generously giving me permission to reprint the two co-authored chapters (Chapters 2 and 3) in this collection. Particular acknowledgements are made, where appropriate, in each chapter, so this is the place to note more general and diffuse debts. But, as Fiona Cowie would say (indeed, has said) in this situation, first, spleen.

I have always been interested in the connections between philosophy and the natural sciences. For most of my professional life, I have thought that the "linguistic turn" in philosophy has been a catastrophe for my discipline. According to this mistake, there is a sharp distinction between empirical and conceptual truths, and the aim of philosophy is to discover purely conceptual truths: truths describing a concept's structure. These discoveries were originally expected to be in the form of a set of necessary and sufficient conditions for the application of philosophically important concepts, though this has proved to be a Moscow from which analytic philosophy was forced to retreat. They were to be discovered by thought experiment: the description (typically, a half or quarter-description) of often physically impossible scenarios about which an intuitive judgement would be elicited. That judgement, the idea ran, would directly reflect the real underlying structure of our concepts. So intuitive judgements about a range of actual and possible scenarios would test proposed analyses. In my view, this whole project has nothing to recommend it. Even if there prove to be conceptual truths, it is far from obvious that their discovery is a worthwhile project. Even if concepts are logically structured, it is far from obvious that our intuitive judgements reflect that structure. And it turns out – I think particularly clearly in philosophy of biology and evolutionary theory – that the empirical/conceptual distinction is no sharp divide. Moreover, the project as a whole has led to a divorce between philosophy and the natural sciences. So goodbye to all that. Instead, we should think of philosophy as part of the same project as the natural sciences, but a discipline concerned with the more theoretically obscure elements of those sciences (this is the grain of truth in the analytic project) and as playing an integrative role within them. I intend these essays to vindicate that metaphilosophy.

I have never defended analytic philosophy (in the narrow sense), but I first pursued this naturalistic project in philosophy of language and philosophy of mind. In the early 1980s, however, I became a born-again Darwinian, first becoming interested in philosophy of biology, and then coming gradually to focus on it. Originally, this was under the influence of a then-student and now-colleague, Peter Godfrey-Smith. He urged me to read *The Selfish Gene* and *The Extended Phenotype*, rightly claiming that they were philosophically as well as empirically rich. Amusingly, given his later Lewontinization, he at first rejected the moderately Dawkinsish *Return of the Gene* as a sell-out. I have talked philosophy of biology; biological philosophy of mind; and biology with him ever since, and we have regularly exchanged manuscripts and ideas. Intellectually, then, this collection owes more to him than to anyone else. At about the same time, Paul Griffiths arrived at the Australian National University to begin a Ph.D. His initial interests were in philosophy of mind, especially the emotions, but we both began to drift towards evolutionary theory at about the same time, and over the following fifteen years I have learnt much from Paul.

Having begun to work my way into the field, I was given much help and encouragement by Elliot Sober and, a little later, by Philip Kitcher. I met David Hull for the first time at an APA conference in 1990, and like almost everyone in the field I have been overwhelmed by his intellectual and personal generosity. I do not think he has ever failed to read and comment on a word I have sent him. He organised a substantial visit to Chicago in 1993 (and insisted on me infesting his house for three months); he encouraged, supported, and edited my recent book with Paul Griffiths. He's a wonder.

In 1988 I moved from the Australian National University to Victoria University of Wellington, New Zealand ("Vic" to its inmates), though I have recently returned half-time to Canberra, and have completed this project while in residence there. My New Zealand base has supported my evolutionary immersion in many ways. I have had the opportunity to teach advanced courses on this material on a regular basis. So many of these ideas were first tried out, in a crude and inchoate form, on generations of Wellington students. Many of these students were a real pleasure to teach: enthusiastic, generous, and able. But I would particularly like to mention in this connection James Maclaurin, now at Otago University; Matthew Barrett, now at Stanford; Roger Sansom, now at Chapel Hill; and Mike Dickison, now at Duke: actual and potential members of a new generation of philosophers of biology. My department, and through it the university, supported my work in many other ways as well. They provided me with time, space, and occasionally the concrete help of research assistants. Most important, it is a supportive, convivial, civilised, and committed place at which to work. My away base, the Philosophy Program in the Research School of the Social Sciences, too, is a wonderful place to work; a heaven away from home.

Philosophy of biology is necessarily interdisciplinary. Science departments do not always welcome philosophers, and knowing the philosophers I do, that is hardly surprising. But that is certainly not true of Vic. The Schools of Biological and Earth Sciences have both been very hospitable, making me more than welcome at their seminar series. In this regard I would like to particularly mention the evolutionary theory gang of four, Mike Hannah (Earth Sciences) and Charlie Daugherty, Geoff Chambers, and Phil Garnock-Jones (Biological Sciences). The three supremos of the Friday Evolution and Ecology Seminar Series, Kath Dickinson, Don Drake, and Christa Mulder deserve my thanks, too (though perhaps not that of my audiences), for inviting me to speak at these seminars, and forcing others to listen. I should also thank the previous publishers of nine of the essays for allowing me to reprint them. In doing so, I have resisted the considerable temptation to alter and update them, but typos, when noticed, have been corrected, and spelling and citation style have been made uniform.

Finally, on a personal note, I need to thank my partner, Melanie Nolan. Research and writing are very hungry of time and energy. So even when I am at home base in Wellington, she is often short-changed. When I escape to Canberra, she gets no change at all. So I owe much to her patience, tolerance, and (even) good humour. That is especially so, since she herself has recently launched on the major research project of motherhood.

Kim Sterelny

Philosophy, Research School of the Social Sciences, Australian National University, and Philosophy, Victoria University of Wellington

Part I

Overview

1

Evolution and Agency

A User's Guide

I OVERVIEW

Earlier work in philosophy of biology focused mostly on understanding the relationship between biology and the more basic sciences
of physics and chemistry, and developing more subtle views on the
major themes of philosophy of science – reduction, explanation and
causation – based on a broader range of scientific practice. Those
issues are still alive in the literature (see, for example, Dupre 1993;
Rosenberg 1994), but in my view philosophy of biology acquired
its contemporary character when the central debates shifted to issues
internal to biology itself. The first such issue was the debate within
evolutionary theory about the nature of selection. In his classic
reconstruction of the Darwinian program, Mayr isolated five independent evolutionary theses: evolution has occurred; contemporary
species have ultimately descended from at most a few earlier life
forms; evolutionary change is typically gradual; species normally
form when lineages split and the fragments diverge, and the
mechanism of adaptive evolution is natural selection (Mayr 1991).
Many of these ideas are no longer controversial. No-one within the
community doubts the fact of evolution, nor that contemporary
life is descendant from a single ancestor (or perhaps a few).
Almost no-one doubts that natural selection has played some
significant role in this process. But there continues to be much
debate on the nature of that role and its relation to species and speciation. The recent boom in philosophy of biology (in part caused, in
part signalled, by Sober [1984]) began with the attempt to understand and resolve these controversies. They are central to this
work, too.

3

Views about selection have tended to cluster into two camps; I will tendentiously call these the American and the British tendencies. The Americans have tended to suppose:

1. The most central phenomenon evolutionary biology must explain is diversity and constraints on diversity.
2. Selection is important, but the evolutionary trajectory of populations is affected by much more than within-population natural selection.
3. When selection acts, it typically acts on individual organisms. Individual organisms within a population are more or less fit. Those fitness differences may not result in differential reproduction (for actual fitness can vary from expected fitness), still less to an evolutionary change in the population as a whole. But the primary bearers of fitness differences are individual organisms.
4. Theorists within this group are often pluralists. That is, they think that while the usual bearers of fitness properties are organisms, they are not the only biological individuals who are more or less fit. Groups of organisms, and even species, are potentially units of selection. In special circumstances even individual genes are units of selection. In particular, this is true of so-called outlaw genes; genes that improve their own chance of replication at the expense of the organisms in which they reside. Such a gene, according to this tendency, really is a unit of selection. But these genes are the exception rather than the rule.
5. The American school is typically sceptical of attempts to apply evolutionary theory to humans. They do not suppose that there is an "in-principle" problem in studying human evolution. For the members of this tendency agree that we came into existence as a standard product of evolution. But they are sceptical of actual attempts to apply evolutionary theory to human social behaviour; in particular they are sceptical of sociobiology and its intellectual descendants.

Paradigm Americans are Lewontin, Gould, Levins, Sober, and Lloyd (oddly enough, all Americans). In the British tendency, we find exemplified the following ideas:

1. The fit between organisms and environment – adaptedness or good design – is the central problem evolutionary biology must explain.
2. Consequently, though the British tendency does not deny the importance of historical, developmental and chance factors in deter-

mining the evolutionary trajectories of populations, natural selection plays a uniquely important role in evolutionary explanation.

3. In some fundamental sense to be explained, the gene is the real unit of selection. This is true not just in the exotic case of sex ratio genes and other outlaws, but in the routine cases as well. For example, a gene that improves the camouflage pattern of a bittern is no outlaw. In improving its own replication prospects, it improves those of all other genes in the genome. But it, rather than the bittern itself, is the real unit of selection.

4. This group is sceptical about "higher order selection"; they doubt that groups of organisms or species form units subject to selection and evolution within competing metapopulations.

5. Members of this group endorse the application of evolutionary theory to human social behaviour, not just in theory but in practice.

Paradigms within the British tendency are Williams, Maynard Smith, Dawkins, Dennett, and Cronin (by no means all British). Indeed, my characterisation of these groups is tendentious, not just in the pseudo-national labels but also in suppressing the fact that these thinkers are less uniform and more finessed in their conceptions of evolution than my ideal typology suggests. But even so, there is a real clustering in positions, and one message of this collection is that this clustering is due to historical accident rather than a necessary connection between these theses. In my view:

1. Both adaptation and diversity are real phenomena of great significance. Both need to be explained. No-one disagrees explicitly with this claim, but in practice evolutionary theories tend to focus on one phenomenon or the other.

2. Natural selection does play a unique role within evolutionary history, a role that must be recognised by evolutionary theory. But selection does not, in general, consist in the adaptation of organisms to their environments.

3. There is an important sense in which genes are the units of selection. But so are other gene-like replicators. Moreover, the pluralists are also right to insist that groups and species sometimes play a role in evolution strikingly like that played by organisms. The members of the British tendency are usually willing to admit the in-principle possibility of such phenomena, but they have been notably unwilling to countenance their actual existence.

5

4. American caution about the application of evolutionary theory to human behaviour is well-founded. But I will try to show by actual example the virtues of an evolutionary perspective on cognitive phenomena.

The next six chapters in this collection explore general themes within evolutionary theory, themes that reappear in a more concrete setting in the final five chapters on the evolution of cognition.

II   THE UNITS OF SELECTION REVISITED; OR WHATEVER HAPPENED TO GENE SELECTION?

Famously, G. C. Williams and Richard Dawkins have argued that the unit of selection is the gene: The history of life is the history of more or less successful gene lineages. On this view, it's a mistake to suppose that organisms are the units of selection. The critical consideration is that evolutionary change depends on cumulative selection. So the units of selection must persist: they must face the tribunal of experience repeatedly, not just once. Organisms are temporary beings; here today; gone tomorrow. They reproduce but are not copied. That is obviously so (the idea runs) in the case of sexual organisms. For no offspring's genome is identical to that of any parent. But asexual reproduction is not copying either. As Dawkins points out, an adventitious change to the phenotype of an asexual organism is not transmitted to its descendants (Dawkins 1982, pp. 97–8). Gene tokens are equally impermanent. But genes, unlike organisms, are copied. Gene lineages – chains of identical gene tokens – are around both today and tomorrow, to repeatedly experience selection's whip.

So when we think of the tree of life, we should think not so much of organisms as gene lineages. Since organisms cannot be copied, they cannot form chains in which each link is a copy of the one before it. But since genes can be copied, they can form lineages; chains of copies, with each link being a copy of its predecessor. The gene's-eye view of evolution takes over the notion of competition, but applies it to competing gene lineages. For lineages can sometimes be many copy-generations deep and they can vary in bushiness, too. A gene may be copied many times, and the copies may form an increasingly broad lineage as well as a deep one. Alternatively, a gene lineage may be thin, with only a few copies existing at each generation. Differences in copy number are usually not accidental, so genes have properties that influence their propensity to be replicated. These properties are targets of selection. Furthermore, since genes are replicated via the reproductive success of organisms, success for one lineage has implications for others. The life or death of an organism has its evolutionary consequences indirectly, by influencing the copying success of the genes within it. Well-built organisms mediate more effective replication of the genes within them. Selection acts through organisms to target some genes rather than others.

This whole conception has been extremely controversial. But some order was injected into this debate through the replicator/interactor distinction (Hull 1981).[1] Obviously, it would be mad to deny that organisms play some especially significant role in evolutionary processes. Gene selectionists have tried to accommodate this obvious truth by distinguishing two different roles in evolution. Replicators transmit similarity across the generations, and interactors interact with the environment with varying success, hence biasing the transmission of replicators. For replicators help construct those interactors, and hence replicators that construct successful, well-adapted interactors do better than those implicated in the construction of less successful ones. Organisms, naturally, are paradigm interactors. In finding a role for organisms, the replicator/interactor distinction resolved some of the problems in the gene selection debate. But very tough questions remain.

First, there is a threat of triviality. Once gene selectionists develop the replicator/interactor distinction, it may be that their theory merely renames the familiar distinction between genotypes and phenotypes. Can gene selectionists stake out some distinctive territory while still recognising the complexity and indirectness of gene's causal actions in the world? Second, gene selection ultimately depends (or so I will argue) on the informational conception of the genome; a conception which has recently come under sustained criticism. Third, it is one thing to develop and defend a distinction between replication and interaction; it is another to show that the genes are the replicators, and organisms the interactors. Replicator selection may not be gene selection, and organisms may not be the only interactors. Let me sketch these ideas a little more fully, and indicate the places I take them up later in this collection.

Is Gene Selection a Genuine
Alternative to the Standard View?

Once gene selectionism comes to rely on the replicator/interactor distinction, perhaps that view just renames the elephant, for the genotype/phenotype distinction is standard fare in evolutionary biology. So, too, is the assignment of fitness values to individual gene types by calculating the average fitness of the bearers of an allele in a population. Calculating gene fitness is a standard way of tracking or representing evolutionary change. So much so, that evolution itself has occasionally been defined as change in gene frequency. How does gene selection, shorn of its metaphors, differ from this banal conception of evolution? Elliot Sober, in particular, has pressed this objection. In his view, gene selectionists face a dilemma: They are pushed either into empirically unsupportable views about the nature of the genotype/phenotype relation or into a view that is nothing more than a restatement of the received conception (Sober 1993).

I argue in Chapter 2 and Chapter 3 that this very deflationary view of gene selection would be a mistake, for it overlooks the significance of Dawkins' ideas about extended phenotypes. Genes have phenotypic effects that influence their replication propensities. If those effects were almost always effects on the design of the organisms carrying the gene, gene selection would rename the phenotype/genotype distinction. But some genes have short arms, and others have long ones. Outlaws typically have short arms. For example, the critical effect of a segregation distorting gene is on its rival allele. The existence of outlaw genes is not controversial; neither is the idea that they are genuine cases of gene selection. It is worth remarking, though, that this concession is less innocent than critics of gene selection suppose. One standard argument against gene selection is that genotype/phenotype relations are too variable and indirect for genes to be "visible" to selection via their phenotypic effects. But it is very hard to claim that *only* outlaw genes are selected in virtue of their fitness properties, for what are they fitter than, in virtue of their being outlaws? Presumably, the normal, co-operating, phenotype-building genes. But if we can think of these ordinary genes as being "visible" to selection (and selected against) in the presence of outlaws, it is hard to see how they could cease to be visible when there are no outlaws. Arguments against gene selection that depend on the complexity and indirectness of the effects of genes on phenotypes threaten to unravel, once the example of outlaws is conceded to gene selection.[2]

The existence of long-armed genes – genes with extended phenotypes – is more controversial. But if Dawkins is right, they are more widespread. There is a parasitic barnacle from the *Rhizocephala* group that chemically castrates and feminises her host, a crab. The genes responsible for that effect replicate in virtue of effects outside the body of the parasite. These genes are not outlaws. They promote the replication of every gene in the parasite genome. But their critical adaptive upshot is not a feature of the parasite's body. Outlaws may be exceptional. But extended phenotype effects are widespread in nature. Parasites themselves are exceptionally numerous, and virtually all carry genes that manipulate hosts and/or suppress their defences. Moreover, many other organisms – organisms whose way of life is not parasitic – carry manipulation genes. As Krebs and Dawkins point out, much animal signalling is an attempt at manipulation (Krebs and Dawkins 1984). Furthermore, mounds, nests, tunnels, casings and the like are adapted structures but are not part of the phenotype of individual organisms. Their adaptations, too, are the result of extended phenotypic effects. Gene selection, then, directs our attention not just to outlaws but to these effects; important evolutionary phenomena we might easily overlook if the standard model were our conception of evolution. It is no mere relabelling of the genotype/phenotype distinction.

The Informational Gene

Why suppose genes form lineages of copies, whereas organisms do not? It is true, of course, that while organisms resemble their parents in many respects, there are differences as well. In particular, the genotype of those organisms formed through sexual reproduction will be like that of neither parent (though, of course, a defence of gene selection must show rather than presuppose the special significance of gene similarities). So perhaps the critical difference is fidelity. When a gene is copied, normally the copy is exactly the same as the template, and that is never true of reproduction. But that idea leads to problems. When all goes well, gene replication does produce new genes that share their antecedent's base sequences. But the route from base sequence to phenotypic effect is complex, indirect and many-many, for in different cellular contexts the same base sequence can yield a different effect, and vice versa. Hence, the biologically relevant properties of a gene token include much more than its base sequence. The surrounding genes and other aspects of the cellular context will play a role in determining whether and how a sequence is transcribed. Hence, organism lineages cannot clearly be distinguished from gene lineages on grounds of copy fidelity.

In my view, the informational conception of the genome is one foundation for the view that gene lineages have a special status. An organism's genome (on this view) is a set of instructions for making that organism. Gene copying is the ground of reproduction, for it's the mechanism through which the instructions for making organisms are transmitted over time. This view is very widely expressed. Here is one recent instance:

> How is it . . . that an egg develops into a mouse, or an elephant or a fruitfly, according to the species that produced it? The short answer is that each egg contains, in its genes, a set of instructions for making the appropriate adult. Of course the egg must be in a suitable environment, and there are structures in the egg needed to interpret the genetic instructions, but it is the information contained in the genes that specifies the adult form. (Maynard Smith and Szathmary 1999, p. 2)

If this view of genes and the genome could be supported, it really would vindicate gene selection. It would suggest that, most fundamentally, evolution is a change in the instruction set. But to put it mildly, this idea is controversial. In particular, Developmental Systems Theorists argue that there is no sense in which only genes carry information about phenotypes. In their view, the only theoretically innocent view of information is covariation. It is true that if we hold the rest of the causal context constant, particular alleles covary with phenotypic outcomes. But that is true of other developmental resources. For example, many reptiles have temperature dependent sex determination. So for them, a particular incubation temperature, holding other factors constant, covaries with a particular sex. So by the same logic, we should say that a particular temperature codes for, programs, or carries the information that the crocodile is female. Developmental Systems Theorists regard this as a reductio of informational conceptions of the role of the gene in evolution.

When we consider the relationship between generations in an evolving population, we are bound to notice that there are many causal connections between one generation and the next, and the similarity between generations – the heritability necessary for natural selection – is the result of many causal factors, not just the flow of genes across the generations. For example, many species of seabird show fidelity to their nest sites, so the mother's choice of site determines, in all probability, that of her offspring. The fundamental challenge of Developmental Systems Theory, then, is their demand for a reason for treating one of these similarity-maintaining causal channels as special. The idea that the information to reconstruct the next generation flows only through the genes is one way of meeting that challenge. But it is not the only way. Dawkins has argued that the genetic channel, and only the genetic channel, meets the replicator condition. If there is genetic change between, say, the parental and the F1 generation, that change will be transmitted to the F2 generation and beyond. The replicator condition, in turn, is necessary for cumulative selection, for only if the mechanisms of inheritance meet the replicator condition will a chance improvement be preserved as the potential basis for further improvement (Dawkins 1982).

I agree that this condition is important, but in Sterelny (2000b) I argue that Dawkins' replicator condition is a special case of a set of conditions on evolvability, and that genetically mediated inheritance meets evolvability conditions better than other inheritance channels. But I also argue that other mechanisms meet evolvability conditions to a significant degree, so that contrast between genetic inheritance and some forms of non-genetic inheritance is only one of degree.

Developmental System Theory's challenge to the concept of genetic information raises difficult and important issues; it is the subject of much ongoing debate.[3] It is the focus of *The Extended Replicator*, in which I argue: (a) covariation is not the only theoretically innocent account of informational notions; biological function is another; (b) given this alternative foundation, genes do (and temperatures typically do not) represent phenotypic outcomes on which they are targeted; and (c) so do some other developmental resources. This debate has been carried further in Maynard Smith (forthcoming) and the commentaries on it. This leads to the next tough question for gene selection.

Identifying Replicators and
Interactors

Let's suppose that the replicator/interactor distinction is sound. Are genes near enough the only replicators? Are organisms near enough the only interactors?

First, replicators. There is a pronounced shift in position from *Return of the Gene* to *The Extended Replicator*. *Return of the Gene* grossly underestimates the extent and the significance of non-genetic inheritance. Genes are far from being the only developmental resources that are both adapted for specific developmental outcomes and transmitted reliably across the generations. Cultural inheritance, broadly conceived to include song and other species-specific calls, nests sites and the like, is a widely accepted supplement to genetic inheritance. In my view, the transmission of symbiotic associates in reproduction is an even more striking phenomenon. In many species there are complex adaptations to ensure the faithful replication and transmission of symbiotic micro-organisms from one generation to the next. If the information needed to build organisms is transmitted across the generations, the instruction set includes non-genetic replicators.

Second, interactors. Defenders of gene selection have been notoriously sceptical about "high level" selection. They have had grave doubts about the idea that groups or species constitute metapopulations characterised by inheritance, variability and differential success and hence undergoing selection. Indeed, G. C. Williams first explicitly formulated gene selection in the course of responding to group selection (Williams 1966). However, once the replicator/interactor distinction was formulated, it has become clear to everyone that there is no conflict between gene selection and group selection. Defenders of group selection argue that groups are organism-like. The key claim is that groups are interactors (Sober and Wilson 1998; Wilson and Sober 1994).

Gene selectionists qua gene selectionists can certainly allow this. By their lights, if group selection exists, it is a special case of an extended phenotypic effect. Organisms would carry genes whose adaptive phenotypic effect is expressed at the level of social organisation, perhaps promoting a certain kind of co-operative interaction. The gene would be replicated in virtue of that effect. However, despite the formal consistency of gene and group selection, members of the gene selection

12

camp have continued to express great scepticism about group selection; see, for example Dawkins (1994).

In my view this is a mistake, and I argue to that effect in Chapter 4. The conceptual and empirical problems surrounding group selection are very thorny, but D. S. Wilson has made a powerful prima facie case for its importance. In his model of group selection, he envisages divided populations. A Wilson population is structured into temporary associations which eventually dissolve back into the population as a whole; a population which then redivides into new temporary associations. Consider, for example, insects that feed on an abundant but rare and widely scattered resource. An instance might be blowflies feeding on kangaroo corpses in an Australian woodland. A species with this way of life will have a divided population structure. The individuals will exist in temporally isolated local concentrations. But as the resource is used up, these temporary aggregations merge back into the larger population. The final generation of maggots must look elsewhere after they turn into flies.

Wilson points out that a population structure of this kind opens the door to group selection. If these local associations differ, and in ways that result in a difference in productivity (i.e., in the number of new organisms they inject into the population as a whole) and if the groups that re-form tend to inherit the characteristics of the groups that produced their constituent organisms, then group selection can take place. Sober and Wilson (1998) argue that this mechanism explains the many observed departures from 50/50 sex ratios in arthropods, for these often have population structures of this kind. An environment in which a key resource is locally abundant but globally rare and scattered selects for female-biased sex ratios at the group level, for when the resource runs out, most individuals are doomed. But a single mated female in making it to a kangaroo corpse can found a new group. So female-biased groups are likely to inject more individuals into the next generation. And many such female-biased sex ratios have been found in nature. They develop a similar model to explain the evolution of less virulent strains of disease. Individual selection on bacteria and other disease organisms typically selects for virulence, but group selection usually selects for less virulence. They argue observed levels of virulence suggest both modes of selection are in operation.

The Wilson model raises difficult issues. It's easy to accept the intuitive soundness of group selection when the groups in question are ant nests, termite mounds, or colonial marine invertebrates. These are enduring, integrated, obviously co-adapted collectives. But Wilson and Sober extend their model way beyond these cases to include much more temporary and/or diffuse aggregations of organisms. For them, any coalition of co-operating baboons, say, constitutes a group, an inter-actor, even if the co-operative episode in question lasts only ten seconds and that is the only co-operative interaction between the animals in their entire lives. Kin selection and reciprocal altruism are typically seen as rival individualist explanations of altruism, but they see them both as special cases of group selection. Hence, their analysis raises in a very sharp form the question: What is it to be an inter-actor? It even raises the question of whether there is a fact of the matter about our count of interactors. Perhaps the question "Is a termite mound an interactor?" has no objective answer. Rather, when it is convenient to do so, we just "take the interactor stance" towards certain evolutionary processes. In Chapter 4, I begin the project of developing a theoretically robust account of the nature of interactors and argue that there are cases of genuine collective interactors – real superorganisms. But I do not think that all the cases that Wilson and Sober treat as group selection fit that picture. Hence, their cases of group selection do not form a cohesive set.

Species Selection

Species selection raises rather different issues for two reasons. First, the species selection literature (with the exception of Williams [1992]) has not been framed with the help of the replicator/interactor distinction, so the hypotheses themselves have often been ambiguous. This is espe-cially problematic because the very nature of species is a matter of such deep controversy within evolutionary biology. Second, while group selection and individual selection offer rival explanations of the same traits – co-operative, apparently altruistic behaviours – that is typically not true of species selection hypotheses. With the partial exception of the evolution of sex, species selection is intended to explain charac-teristics of entire species or lineages, not conventional phenotypic traits. Species selection is typically taken to explain such phenomena as the evolvability of a lineage; its species richness; its distribution over niches. So species selection hypotheses give rise to conceptually diffi-cult problems including:

(a) What view of the nature of species do species selection hypotheses imply? Species selection, I argue in Chapter 5, is committed to a quite rich concept of species, and more.

(b) What counts as a species level trait? For example, does the geographical distribution of a species count as a bona fide trait?

(c) How can we distinguish, empirically and conceptually, between true species selection and "species sorting" (Vrba 1984c).

Species sorting is a lineage level pattern that is a by-product of evolution at the level of individuals in a population. The eucalyptus lineage in Australia is species rich, and geographically and ecologically widespread. It is quite possible that this is the consequence of selection on individual plants in ancestral species for drought and fire resistance some millions of years ago. These hardy eucalypts spread and speciated, giving rise to the diversity of contemporary forms. If this were the right story of Australian eucalypt history, this rich lineage would be a by-product of interaction and replication of individual eucalypt trees and individual eucalypt genes. The lineage level pattern would be an effect of selection on individual organisms. I do not think there are clear solutions to the problems of distinguishing between individual and lineage properties, and the associated distinction between species sorting and species selection. But I do argue that range and evolvability, for example, are lineage level properties (despite, of course, supervening on the properties of individual organisms) and that the idea that they evolve by species selection is at least a plausible conjecture. So Chapters 5 and 6 together offer a tentative defence of species selection. However, the first of these chapters now strikes me as too cautious on this issue, understating the prospects for finding causally relevant species level properties heritable over lineage splits. So I now see that chapter as making the case that the hypothesis of punctuated equilibrium, if sound, establishes that one important condition of species selection is met: species are identifiable, objective units of evolutionary history. And it also shows that the causal importance of species and speciation to evolution does not depend on the truth of a species selection hypothesis. (On this, see also Sterelny 1999b.)

Jointly, then, on the units of selection problem, these chapters defend the following ideas: (i) I regard the replicator/interactor distinction as a good general way of framing questions about units of selection; (ii) I defend a much modified version of gene selection.

15

Genes and other gene-like developmental resources form more or less successful replicator lineages; and (iii) I defend, though in a rather hedged and empirically cautious way, high level selection; both group selection and species selection. In both cases, I take the high level units to be interactors.[4] But I also argue that there are important open questions about the identification of interactors and their phenotypic characteristics.

III NATURAL SELECTION

Adaptationism has been a controversial topic within evolutionary biology for the last twenty years or so. Some of that controversy, though by no means all, eases when we realise that adaptationism is not a single doctrine, but a cluster of loosely related views. In a recent paper, Peter Godfrey-Smith distinguished three different adaptationist ideas, and his taxonomy is a useful way of framing this section (Godfrey-Smith 1999).

Empirical adaptationism is the doctrine that natural selection has been the most important force driving evolutionary history. *Explanatory adaptationism* is the idea that adaptation, the appearance of design, is the phenomenon in biology in most need of explanation, and that natural selection provides that explanation. Explanatory adaptationism could be true and empirical adaptationism false, and vice versa. Explanatory adaptationism is compatible with selection being highly constrained, with its being dominated by drift, or by genetic and historical constraints. Suppose that such constraints impose very great limitations on the evolutionary potentialities of, say, the tree kangaroo lineage. Even so, those adaptations for arboreal life found in tree kangaroos are consequences of selection. Empirical adaptationism could be true while explanatory adaptationism false. For explanatory adaptationism, in taking natural selection to be an explanation of design-like features of organisms, presupposes that we can identify adaptation independently of natural selection. This supposition is controversial. In most contemporary writings, an adaptation is by definition the product of natural selection. *Methodological adaptationism* is not an empirical claim about the history of life at all. It recommends a research procedure. Work out what the phenotype of a population would be like, if natural selection were unconstrained by development, genetic variation and the like. Then compare this optimum phenotype to the actual

16

phenotype. Mismatch reveals the existence of constraints on selection; constraints that are not manifest in phenotypes. Since this method makes no claim about the role of natural selection in evolution, it is independent of the other two versions of the idea.

I have a good deal of time for all three versions of adaptationism, though the methodological thesis does not get much of a run in these pages. Explanatory adaptationism is explored by example. The last part of the collection is a pursuit of adaptationist explanations of cognition. Empirical adaptationism does get extended treatment, for there is a serious problem in unpacking this idea as well as in evaluating it, for everyone agrees that selection itself never suffices for any evolutionary change. Selection needs variation, and the history of the lineage determines both the developmental mechanisms and gene pool of a species, and hence controls the supply of variation. So what could it mean to claim explanatory priority for selection? How can one necessary but not sufficient condition of some evolutionary change be more important than other necessary but not sufficient conditions?

I think this question can be answered in a way that leaves empirical adaptationism a plausible conjecture. In doing so though, it is important not to saddle adaptationism with unnecessary commitments. Godfrey-Smith (1996) treats adaptationism as a species of externalist explanation. In his view, an adaptationist explanation of (say) the stiffened, counter-weighted tail of the tree kangaroo explains a feature of that lineage's phenotype by appeal to the environment in which the lineage is placed. Natural selection shapes organisms to fit their environment; their niche. Natural selection enables a lineage to track and respond to changes in its environment. So selective explanations are "outside-in" explanations.

I have no doubt both that some adaptationist explanations do fit this picture and that overall it's a bad picture of evolution and of evolutionary change. While some episodes in evolutionary history are properly represented as the accommodation of a lineage to an invariant environment – streamlining in dolphins, sharks, and marine reptiles are obvious examples – much selection involves reciprocal change. For the relevant aspects of the environment are often biological; predators, prey, competitors, and parasites themselves change in evolutionary time, often in response to others' changes. Moreover, Lewontin has convincingly argued that organisms frequently both select and modify their environments. The problems of this externalist conception of

adaptationism, and the consequences for both evolutionary and ecological theory of rejecting it, are explored in Chapter 7.

So it would certainly be a mistake to suppose that in general selection involves a causal arrow from environment to lineage but not from
lineage to environment. But I argue in Chapter 8 that adaptationism is
not committed to this misconception. That is well and good, but I still
need to sketch how one could give causal priority to selection whilst
recognising that selection is never in itself sufficient for any evolutionary change. The first step, developed and defended in *Explanatory
Pluralism in Evolutionary Biology*, is to distinguish between two types
of explanation: robust process and actual sequence explanation.

It's helpful to introduce this distinction with a non-biological
example. There was a dramatic fall in birth-rate in New Zealand after
World War I. How might we explain that change? One idea, possible
in principle but rarely in practice, is to assemble the biographies of the
agents in question, and show how their individual reproductive careers
emerged out of their intentions, decisions, and actions. The demographic pattern can then be derived as an aggregate of these individual facts. There are micro-histories that attempt causal narratives of
this kind on a small scale. Obviously it's impossible on a national scale,
even for a small country like New Zealand. A second idea is to argue
that the demographic change is a consequence of urbanisation. New
Zealand, like many other Western countries, saw a population shift
from the country to the larger towns. Urbanisation changes the costs
and benefits of children, so birth-rates declined.

In one way the first explanation is more information rich. It gives us
to any level of precision we like the actual chain of events underlying
the effect we seek to explain. But Frank Jackson and Philip Pettit have
pointed out that explanations of the second kind are not just vaguer
versions of the first, to be accepted only because of practical limits on
the detail and precision to which we can aspire, for the second tells us
something that the first leaves out, namely, that the effect is robust.
Changes in the initial conditions or causal sequence that do not undermine the general flow from country to town, or the costs and benefits
of urban children, will leave the demographic effect intact. Intuitively,
robust process explanations identify the class of worlds in which we
have the effect: all the urbanising worlds. Actual sequence explanations
identify our world in that class (Jackson and Pettit 1992).

Implicitly in Chapter 2, and explicitly in Chapter 6, I argue that
adaptationist explanations are robust process explanations. Adapta-

tionist explanations of, say, industrial melanism in the peppered moth are committed to the view that a melanic population would have evolved under the selective regime induced by industrial pollution even had the initial conditions been different. Differences in the frequency distribution of genes in the gene pool; the timing of melanic mutations; variations between actual and expected fitness all might have resulted in a different evolutionary trajectory. But a melanistic population would have evolved just so long as the same selective regime was in place. Melanism could have moved to fixation via different trajectories and at different rates. Nonetheless, if the adaptationist hypothesis is right, these are all trajectories to a dark-coloured populations.

The distinction between robust process and actual sequence explanations shows how it's possible to see natural selection as explanatorily fundamental despite the fact that drift, accidents of timing of mutations, and the initial frequencies of genes in a population play a role in every evolutionary change. But what of the basic developmental system of the moth species? After all, it is only due to the lineage's history that melanism is an evolutionary possibility for that lineage. The inherited set of developmental mechanisms determine the lineage's set of evolutionary possibilities. Other moth lineages inheriting a different set of mechanisms respond to changes in predation regimes by behavioural changes or by the development of toxic substances, mimicry and the like. So surely adaptationist explanation, even construed as robust process explanations, in concentrating on selection leave something critical out: the historical determinants of the lineage's evolutionary envelope.

The literature on historical and developmental constraints and their relation to adaptationism is complex. But I have argued elsewhere (Sterelny 1999a; Sterelny forthcoming) that there are three main possibilities we need to consider.

1. It might turn out that in most lineages the variation/selection/variation cycle can generate phenotypic change in any direction, so the mechanisms that underpin phenotypic variation do not significantly bias a lineage's evolutionary envelope. This is a live option. In particular, it has not been refuted by the discovery of widely conserved developmental and genetic resources. One of the most striking aspects of these discoveries is that conserved genetic mechanisms generate a variety of developmental outcomes, often via gene

duplication followed by modification. Suppose, then, that developmental mechanisms do not bias a lineage's evolutionary possibilities. If so, though developmental biology does help explain variability, and though variability is a necessary condition of any evolutionary change, nonetheless development can be treated as a background condition to evolution.

2. Perhaps the mechanisms that underpin phenotypic variation do bias a lineage's evolutionary possibilities, making some changes more likely and others less likely. If this turns out to most often be the case, the evolutionary trajectory of a lineage is explained by both factors external to a lineage and factors internal to it, including factors that bias the generation of variation.

3. Perhaps a proper appreciation of developmental biology does not just contribute to the answers evolutionary biologists seek. It changes their questions. On this view, the most fundamental question that confronts evolutionary biology is: How is the evolution of complex adaptation possible? On this view adaptation is a real phenomenon, and is rightly explained by selection. But it is made possible only by very special developmental scaffolding. The bedrock agenda of evolutionary biology is to provide an explanation of the evolution of that scaffolding.

These alternatives differ in their empirical bets about development. Empirical adaptationists are often supposed to be committed to the first hypothesis: that the history of a lineage does not constrain its evolutionary possibilities. But that is not so. As I have argued in Chapter 8 (and in more detail in Sterelny and Griffiths 1999), empirical adaptationists can live with a version of the second hypothesis. The key idea is that in abstracting away from historical constraints, the adaptationist is supposing that these constraints are stable over evolutionarily significant chunks of time and hence over sizeable chunks of phylogenetic trees. If the evolutionary envelope – the space of evolutionary possibilities – is largely the same for all the moths of the peppered moth family, then *differences* within that family can be reasonably attributed to selection. On all these issues the jury is still out. But empirical adaptation is alive, if either conserved, inherited developmental mechanisms do not significantly bias a lineage's evolutionary envelope, or if those biases are often the same for all species in large clades. For then, too, they can be treated as invariant background conditions.

IV   THE DESCENT OF MIND

What kinds of minds are there, and what are they for? I am adaptationist enough to assume both that these questions are related and that, ultimately, understanding the biological functions of cognition will explain the variety of cognitive engines. So the second question is more causally fundamental than the first. Methodologically, I expect each to throw light on the other. Observed variation in kinds of minds should suggest hypotheses about the biological function of cognition, and hypotheses about biological function should suggest cognitive capacities we might expect to find in the wild. In these chapters I have attempted to develop a very general picture of cognitive evolution, but I also attempt to apply these ideas to a specific evolutionary-cum-architectural hypothesis about hominid evolution.

In these chapters, three themes are central. The first focuses on the evolution of belief-like states, and the evolutionary transition from organisms that detect and respond to their environment in very simple ways to more complex representation by an organism of its environment. When is such representation necessary? In recent literature in the philosophy of psychology, it has been argued that "world models" are typically not necessary for the control of intelligent, adaptive behaviour (Brooks 1991; Clark 1997). I argue that these arguments are flawed through a skewed choice of examples. They focus on adaptive behaviour directed at the inanimate world, rather than response to the biological environment. So the first major theme of these chapters is that the biological world is typically hostile, not indifferent, and that this hostility has *epistemic* consequences for agents. If an agent's response to its biological environment is to be robustly adaptive, that agent requires extra cognitive resources. In particular, it requires the capacity to represent that world. So I argue that the existence of the hostile world of competition and predation is the key selective driver of the evolutionary shift from simple detection of states of the environment to representing the world.

My second overarching theme concerns the evolution of intentional agency. An intentional agent is an agent whose behaviour is explained by both beliefs and preferences, but the role of preferences has been underplayed in evolutionary speculation on these issues. It has been tacitly assumed that in the evolution of intentional agency, belief and preference are a package deal. But there is no a priori reason to suppose that the evolution of belief-like representational states is

21

linked to the evolution of preference-like states. An organism could have belief-like representations of its world while still having a relatively simple motivational system based on a hierarchy of drives. So explaining why animals evolve the capacity to have belief-like states does not automatically explain why preferences have evolved. In these chapters, then, I focus on the issue of preference as much as belief. So I am concerned with the question "what use is a utility function?" Though I develop some answers to that question, I accept that these are speculative, and the smart money will not be on their being right. The critical point, though, is that we need an answer to this question, and, consequently, we also need some idea of evolutionary trajectories from non-representational motivational structures through to a preference structure.

A third theme focuses on a more specific issues: the social intelligence hypothesis and the evolution of "second order" intentional systems. Such agents are able to represent not just their physical environments but also their cognitive environment; the thoughts of others. There has been great interest in the primate evolutionary transition from "behaviour readers" – animals that know about others' behavioural dispositions but not the mental causes of those behaviours – to "mind readers" – primates with beliefs about beliefs. This distinction rests on a false dichotomy. I think it is most likely that real primates, including the great apes, are neither mere behaviour readers nor possessed of something like our folk psychology. And I try to chart out some of these intermediate possibilities. In the process I argue that the cognitive significance of second order intentionality has been much over-sold.

No doubt there are many different ways of taxonomising kinds of minds: I do not expect the one exemplified in these chapters to be exhaustive. It certainly does not draw all the distinctions in which one might reasonably be interested. Nor is it privileged. No doubt one could quite legitimately divide up the pie according to entirely different principles than the one I have chosen. That said, the taxonomy in these chapters enables us to pose important questions about the nature of cognition and its evolution. Let's begin with the basic framework.

Just about all organisms are capable of selective response to some feature of their environment. As is well known, even bacteria do it. Anaerobic bacteria, for example, use magnetosomes to detect the direction of oxygen-free water. So even the simplest creatures exhibit some minimal form of behavioural plasticity in response to signals from their environment. Yet flexible response is not free. There will be metabolic costs in building the machinery that guides flexible response. But there is a more subtle cost, too, the cost of error. The more an organism specialises its behaviour to a particular circumstance, the less adaptive that behaviour will be if the organism has misread its environment. So selection will favour flexible response only if environments vary in ways relevant to the organism; the organism has some means of tracking variation in the environment and matching its behaviour appropriately to that variation; and that optimising behaviour to particular circumstances is worth the cost of error.

Godfrey-Smith explores this idea for very simple forms of plastic response. I have been interested in extending this framework to more complex cases. I have argued in Chapters 9, 11, and 12 for a fundamental distinction between stimulus-bound systems of behavioural control, where adaptive response to the world is mediated by a single proximal stimulus, and multi-tracked behavioural control. Such organisms that can track features of their environment in more than one way, they are equipped with several channels through which information (and on occasion misinformation) flows. Stimulus-bound behaviour can be quite complexly structured. But as Tinbergen showed in his early ethological work on seabirds, those behaviours are fragile. Completely inappropriate behaviours – rejecting a nestling, or attempting to brood a supersize dummy egg – can be induced by manipulating that single stimulus (Tinbergen 1960).

In *Basic Minds* and *Situated Agency and the Descent of Desire*, I argue that the contrast between stimulus-bound response and multi-tracked response is the distinction between mere detection of the world and genuine representation of it. I still think that this claim accords fairly well with our judgements on particular cases, but I have been persuaded that there is an element of stipulation to it. In any case, that conceptual distinction is inessential to the idea that there is a crucial difference between stimulus-bound and multi-tracking organisms, and to an adaptive conjecture about the evolution of that difference. As I noted above, stimulus-bound behaviour is fragile; hence, organisms that can multi-track features of their environment have a more robust capacity to respond adaptively to their world. That capacity is needed, I have suggested, when the organism's environment is translucent. Translucence, in turn, is to be expected when the world is hostile rather than benign or indifferent.

Consider, for example, the problem posed to an animal by the task of walking over rough or uneven ground. Walking is such a natural and everyday activity for us that it's easy to overlook the complexity of the task. That complexity becomes all too obvious to those who have attempted to design insect-like walking robots. Notice, though, that the terrain does not care whether an insect can walk over it or not. Terrain is not selected to actively sabotage or hinder an insect's progress. An animal's predators, prey, and competitors are under selection to sabotage its actions. Adaptive behaviour targeted on the inanimate world (and biologically indifferent parts of the animate world) can rely for control on simple cues of the relevant environmental structure. So the world will often be in those respects *transparent*: The animal can easily see through proximal stimulation to the opportunities its environment offers. It is much more risky to depend on simple cues when interacting with rivals and enemies, since these epistemically pollute the organism's environment. The species-specific mating signals of fireflies are mimicked by predators; reliance on this simple cue is decidedly dangerous. The world, in these respects, is *translucent*. There is information about predators, prey, and competition available to an animal in the proximal signals it can detect. But the relation between proximal signal and the relevant structure of the environment is often complex and noisy.

I think this is an interesting perspective not just on the evolution of representation from simpler adaptive responses but also on the evolution of intentional systems. Most of the empirical and theoretical literature on intentional systems has focused on belief, and even on beliefs about beliefs. For there has been a lively debate on hominid and primate "theories of mind" and on the role of a theory of mind in hominid cognitive evolution. In this literature, the role of desires or preferences has been neglected. I think it has been tacitly assumed that the evolution of belief and of preference is a package deal. In my view, this should not be assumed. There is no uncontroversial way of identifying the features of mental representation that are definitive of beliefs. But I think the following would be on most lists: (i) Beliefs are functionally independent of particular acts. They are relevant to many possible actions, but particular beliefs do not have the function of driving particular behaviours; (ii) Perhaps beliefs must be part of a representational system; a capacity to generate beliefs must be systematic and productive to at least some degree; (iii) Beliefs are true or false. I see no a priori reason to suppose that the evolution of such representation is tied to the liberation of motivation from sensations and drives, and placing it under the control of representations of potential changes to the world. So in Chapters 11 and 12, I explore the evolution of preference. What kind of animals need representationally fancy motivation systems? When do sensations cease to suffice?

I also use this framework for probing an important contemporary debate about cognitive evolution, the social intelligence hypothesis. Of course, there is no single social intelligence hypothesis: it is a generic rather than a specific thesis. One dimension of difference concerns selective forces, and hence what primates or hominids need to know about their social environment. Dunbar and Alexander, for example have emphasised human cognitive adaptations to life in large groups, whereas Byrne and Whiten have emphasised deceptive and counter-deceptive strategies (Alexander 1987; Byrne and Whiten 1988; Dunbar 1996). Another dimension concerns cognitive mechanisms: How do hominids represent and use what they know? But despite these variations, the following two hypotheses are widely accepted:

1. The selective force driving cognitive evolution in the hominid lineage (or perhaps the great ape lineage) is selection for social competence. Competence involves not just anticipating the behaviour of others in the group, but shaping that behaviour and taking advantage of it; sometimes in co-operative interactions and sometimes in competitive ones.
2. The architectural breakthrough – the cognitive Great Leap Forward – induced by selection for social intelligence is metarepresentation. We, and perhaps some of the great apes, do not just think; we think about thoughts. And that, directly or indirectly, is why we are so smart. Social competence of a high order depends on having a "theory of mind," an understanding of something like folk psychology.

I am very sceptical about this second line of thought. That scepticism is not targeted at the social intelligence hypothesis itself. That hypothesis claims that primate fitness depends heavily on social competence – for example, because alliances are very important in those societies and yet these societies are complex. This complexity makes social navigation in such societies demanding, and selects for intelligence, thereby making social navigation even more demanding. So the social intelligence hypothesis builds in a feedback loop that explains the runaway, peacock-tail appearance of human intelligence. This idea is quite plausible, though a long way from proven.

My scepticism turns on the link between the social intelligence hypothesis and metarepresentation. One problem (developed in *Basic Minds*) is that there is no single metarepresentation hypothesis but, rather, a cluster of them. Understanding the metarepresentation hypothesis one way, there is nothing special about metarepresentation, and there is no reason to suppose it is restricted to the fancier primates, or distinctive of special cognitive machinery. Understood another way, as a grip on a theory-like folk psychology, it is special. But then, as I argue in Chapter 10, there is no reason to suppose it's needed for social intelligence. We clearly do have metarepresentational capacities, but it is arguable that not even human social intelligence depends on a theory-like understanding of the minds of others. Chapter 10 defends simulationist accounts of social competence against those who argue that human social skills depend on a sophisticated, theory-like, understanding of other agents. It takes the same line on metarepresentational views of great ape capacities, arguing that their social skills, even if they are as impressive as reported, do not require a close approximation to folk psychology. There are various advanced social skills that have been taken to be diagnostic of metarepresentation: mirror self-recognition; imitation; monitoring and manipulating visual attention; role reversal in joint tasks. But rumours to the contrary notwithstanding, none of these require a theory of mind. So, at least, Chapter 10 argues.

NOTES

1. This distinction is related to, but not quite identical with, Dawkins' distinction between replicator and vehicle (Dawkins 1982). For the difference, see Chapter 4, Section 4.
2. There is a different argument that turns on the presence of outlaws: namely, that in the absence of outlaws, there are no within-genotype fitness differences on which selection can act. But that is a very different argument.
3. See Griffiths, Gray and Oyama: *Cycles of Contingency*; Cambridge, Mass., MIT Press, 2000.
4. In *Punctuated Equilibrium and Macro-evolution*, I note that while one can see a species gene pool as itself a replicator, there seems no explanatory payoff in doing so.

Part II

Replication and Interaction

2

The Return of the Gene

We have two images of natural selection. The orthodox story is told in terms of individuals. More organisms of any given kind are produced than can survive and reproduce to their full potential. Although these organisms are of a kind, they are not identical. Some of the differences among them make a difference to their prospects for survival or reproduction, and hence, on the average, to their actual reproduction. Some of the differences which are relevant to survival and reproduction are (at least partly) heritable. The result is evolution under natural selection, a process in which, barring complications, the average fitness of the organisms within a kind can be expected to increase with time.

There is an alternative story. Richard Dawkins[1] claims that the "unit of selection" is the gene. By this he means not just that the result of selection is (almost always) an increase in frequency of some gene in the gene pool. That is uncontroversial. On Dawkins's conception, we should think of genes as differing with respect to properties that affect their abilities to leave copies of themselves. More genes appear in each generation than can copy themselves up to their full potential. Some of the differences among them make a difference to their prospects for successful copying and hence to the number of actual copies that appear in the next generation. Evolution under natural selection is thus a process in which, barring complication, the average ability of the genes in the gene pool to leave copies of themselves increases with time.

Dawkins's story can be formulated succinctly by introducing some of his terminology. Genes are *replicators* and selection is the struggle among *active germ-line* replicators. Replicators are entities that can be copied. Active replicators are those whose properties

29

influence their chances of being copied. Germ-line replicators are those which have the potential to leave infinitely many descendants. Early in the history of life, coalitions of replicators began to construct *vehicles* through which they spread copies of themselves. Better replicators build better vehicles, and hence are copied more often. Derivatively, the vehicles associated with them become more common too. The orthodox story focuses on the successes of prominent vehicles – individual organisms. Dawkins claims to expose an underlying struggle among the replicators.

We believe that a lot of unnecessary dust has been kicked up in discussing the merits of the two stories. Philosophers have suggested that there are important connections to certain issues in the philosophy of science: reductionism, views on causation and natural kinds, the role of appeals to parsimony. We are unconvinced. Nor do we think that a willingness to talk about selection in Dawkinspeak brings any commitment to the adaptationist claims which Dawkins also holds. After all, adopting a particular perspective on selection is logically independent from claiming that selection is omnipresent in evolution.

In our judgement, the relative worth of the two images turns on two theoretical claims in evolutionary biology.

1. Candidate units of selection must have systematic causal consequences. If Xs are selected for, then X must have a systematic effect on its expected representation in future generations.
2. Dawkins's gene selectionism offers a *more general theory* of evolution. It can also handle those phenomena which are grist to the mill of individual selection, but there are evolutionary phenomena which fit the picture of individual selection ill or not at all, yet which can be accommodated naturally by the gene selection model.

Those sceptical of Dawkins's picture – in particular, Elliott Sober, Richard Lewontin, and Stephen Jay Gould – doubt whether genes can meet the condition demanded in (1). In their view, the phenomena of epigenesis and the extreme sensitivity of the phenotype to gene combinations and environmental effects undercut genic selectionism. Although we believe that these critics have offered valuable insights into the character of sophisticated evolutionary modelling, we shall try to show that these insights do not conflict with Dawkins's story of the workings of natural selection. We shall endeavour to free the thesis of genic selectionism from some of the troublesome excresences which have attached themselves to an interesting story.

The Return of the Gene

I GENE SELECTION AND BEAN-BAG GENETICS

Sober and Lewontin argue against the thesis that all selection is genic selection by contending that many instances of selection do not involve selection for properties of individual alleles (Sober and Lewontin 1982). Stated rather loosely, the claim is that, in some populations, properties of individual alleles are not positive causal factors in the survival and reproductive success of the relevant organisms. Instead of simply resting this claim on an appeal to our intuitive ideas about causality, Sober has recently provided an account of causal discourse which is intended to yield the conclusion he favors, thus rebutting the proposals of those (like Dawkins) who think that properties of individual alleles can be causally efficacious.[2]

The general problem arises because replicators (genes) combine to build vehicles (organisms) and the effect of a gene is critically dependent on the company it keeps. However, recognising the general problem, Dawkins seeks to disentangle the various contributions of the members of the coalition of replicators (the genome). To this end, he offers an analogy with a process of competition among rowers for seats in a boat. The coach may scrutinise the relative times of different teams but the competition can be analysed by investigating the contributions of individual rowers in different contexts (Dawkins 1976, pp. 40–1, pp. 91–2; Dawkins 1982, p. 239).

Sober's Case. At the general level, we are left trading general intuitions and persuasive analogies. But Sober (and, earlier, Sober and Lewontin) attempted to clarify the case through a particular example. Sober argues that *heterozygote superiority* is a phenomenon that cannot be understood from Dawkins's standpoint. We shall discuss Sober's example in detail; our strategy is as follows. We first set out Sober's case: heterozygote superiority cannot be understood as a gene-level phenomenon, because only pairs of genes can be, or fail to be, heterozygous. Yet being heterozygous can be causally salient in the selective process. Against Sober, we first offer an analogy to show that there must be something wrong with his line of thought: from the gene's-eye view, heterozygote superiority is an instance of a standard selective phenomenon, namely *frequency-dependent* selection. The advantage (or disadvantage) of a trait can depend on the frequency of that trait in the relevant population.

Having claimed that there is something wrong with Sober's argument, we then try to say what is wrong. We identify two principles on which the reasoning depends. First is a general claim about causal uniformity. Sober thinks that there can be selection for a property only if that property has a positive uniform effect on reproductive success. Second, and more specifically, in cases where the heterozygote is fitter, the individuals have no uniform causal effect. We shall try to undermine both principles, but the bulk of our criticism will be directed against the first.

Heterozygote superiority occurs when a heterozygote (with genotype *Aa*, say) is fitter than either homozygote (*AA* or *aa*). The classic example is human sickle-cell anemia: homozygotes for the normal allele in African populations produce functional hemoglobin but are vulnerable to malaria, homozygotes for the mutant ("sickling") allele suffer anemia (usually fatal), and heterozygotes avoid anemia while also having resistance to malaria. The effect of each allele varies with context, and the contexts across which variation occurs are causally relevant. Sober writes:

> In this case, the *a* allele does not have a unique causal role. Whether the gene *a* will be a positive or a negative causal factor in the survival and reproductive success of an organism depends on the genetic context. If it is placed next to a copy of *A*, *a* will mean an increase in fitness. If it is placed next to a copy of itself, the gene will mean a decrement in fitness (Sober 1984, p. 303).

The argument against Dawkins expressed here seems to come in two parts. Sober relies on the principle

> (A) There is selection for property *P* only if in all causally relevant background conditions *P* has a positive effect on survival and reproduction.

He also adduces a claim about the particular case of heterozygote superiority.

> (B) Although we can understand the situation by noting that the heterozygote has a uniform effect on survival and reproduction, the property of having the *A* allele and the property of having the *a* allele cannot be seen as having uniform effects on survival and reproduction.

We shall argue that both (A) and (B) are problematic.

Let us start with the obvious reply to Sober's argument. It seems that the heterozygote superiority case is akin to a familiar type of frequency-dependent selection. If the population consists just of *AAs* and a mutation arises, the *a*-allele, then, initially *a* is favoured by selection. Even though it is very bad to be *aa*, *a* alleles are initially likely to turn up in the company of *A* alleles. So they are likely to spread, and, as they spread, they find themselves alongside other *a* alleles, with the consequence that selection tells against them. The scenario is very similar to a story we might tell about interactions among individual organisms. If some animals resolve conflicts by playing hawk and others play dove, then, if a population is initially composed of hawks (and if the costs of bloody battle outweigh the benefits of gaining a single resource), doves will initially be favored by selection.[3] For they will typically interact with hawks, and, despite the fact that their expected gains from these interactions are zero, they will still fare better than their rivals whose expected gains from interactions are negative. But, as doves spread in the population, hawks will meet them more frequently, with the result that the expected payoffs to hawks from interactions will increase. Because they increase more rapidly than the expected payoffs to the doves, there will be a point at which hawks become favoured by selection, so that the incursion of doves into the population is halted.

We believe that the analogy between the case of heterozygote superiority and the hawk-dove case reveals that there is something troublesome about Sober's argument. The challenge is to say exactly what has gone wrong.

Causal Uniformity. Start with principle (A). Sober conceives of selection as a *force*, and he is concerned to make plain the effects of component forces in situations where different forces combine. Thus, he invites us to think of the heterozygote superiority case by analogy with situations in which a physical object remains at rest because equal and opposite forces are exerted on it. Considering the situation only in terms of net forces will conceal the causal structure of the situation. Hence, Sober concludes, our ideas about units of selection should penetrate beyond what occurs on the average, and we should attempt to isolate those properties which positively affect survival and reproduction in every causally relevant context.

Although Sober rejects determinism, principle (A) seems to hanker after something like the uniform association of effects with causes that deterministic accounts of causality provide. We believe that the principle cannot be satisfied without doing violence to ordinary ways of thinking about natural selection, and, once the violence has been exposed, it is not obvious that there is any way to reconstruct ideas about selection that will fit Sober's requirement.

33

Consider *the* example of natural selection, the case of industrial melanism.[4] We are inclined to say that the moths in a Cheshire wood, where lichens on many trees have been destroyed by industrial pollutants, have been subjected to selection pressure and that there has been selection for the property of being melanic. But a moment's reflection should reveal that this description is at odds with Sober's principle. For the wood is divisible into patches, among which are clumps of trees that have been shielded from the effects of industrialisation. Moths who spend most of their lives in these areas are at a disadvantage if they are melanic. Hence, in the population comprising all the moths in the wood, there is no uniform effect on survival and reproduction: in some causally relevant contexts (for moths who have the property of living in regions where most of the trees are contaminated), the trait of being melanic has a positive effect on survival and reproduction, but there are other contexts in which the effect of the trait is negative.

The obvious way to defend principle (A) is to split the population into subpopulations and identify different selection processes as operative in different subgroups. This is a revisionary proposal, for our usual approach to examples of industrial melanism is to take a coarse-grained perspective on the environments, regarding the existence of isolated clumps of uncontaminated trees as a perturbation of the overall selective process. Nonetheless, we might be led to make the revision, not in the interest of honouring a philosophical prejudice, but simply because our general views about selection are consonant with principle (A), so that the reform would bring our treatment of examples into line with our most fundamental beliefs about selection.

In our judgement, a defence of this kind fails for two connected reasons. First, the process of splitting populations may have to continue much further – perhaps even to the extent that we ultimately conceive of individual organisms as making up populations in which a particular type of selection occurs. For, even in contaminated patches, there may be variations in the camouflaging properties of the tree trunks and these variations may combine with propensities of the moths to cause local disadvantages for melanic moths. Second, as many writers have emphasised, evolutionary theory is a statistical theory, not only in its recognition of drift as a factor in evolution but also in its use of fitness coefficients to represent the expected survivorship and reproductive success of organisms. The envisaged splitting of populations to discover some partition in which principle (A) can be maintained is at odds with the strategy of abstracting from the thousand natural shocks that

34

organisms in natural populations are heir to. In principle, we could relate the biography of each organism in the population, explaining in full detail how it developed, reproduced, and survived, just as we could track the motion of each molecule of a sample of gas. But evolutionary theory, like statistical mechanics, has no use for such a fine grain of description: the aim is to make clear the central tendencies in the history of evolving population, and, to this end, the strategy of averaging, which Sober decries, is entirely appropriate. We conclude that there is no basis for any revision that would eliminate those descriptions which run counter to principle (A).

At this point, we can respond to the complaints about the gene's eye view representation of cases of heterozygote superiority. Just as we can give sense to the idea that the trait of being melanic has a unique environment-dependent effect on survival and reproduction, so too we can explicate the view that a property of alleles, to wit, the property of directing the formation of a particular kind of hemoglobin, has a unique environment-dependent effect on survival and reproduction. The alleles form parts of one another's environments, and, in an environment in which a copy of the *A* allele is present, the typical trait of the *S* allele (namely, directing the formation of deviant hemoglobin) will usually have a positive effect on the chances that copies of that allele will be left in the next generation. (Notice that the effect will not be invariable, for there are other parts of the genomic environment which could wreak havoc with it.) If someone protests that the incorporation of alleles as themselves part of the environment is suspect, then the immediate rejoinder is that, in cases of behavioural interactions, we are compelled to treat organisms as parts of one another's environments.[5] The effects of playing hawk depend on the nature of the environment, specifically on the frequency of doves in the vicinity.[6]

The Causal Powers of Alleles. We have tried to develop our complaints about principle (A) into a positive account of how cases of heterozygote superiority might look from the gene's eye view. We now want to focus more briefly on (B). Is it impossible to reinterpret the examples of heterozygote superiority so as to ascribe uniform effects on survival and reproduction to allelic properties? The first point to note is that Sober's approach formulates the Dawkinsian point of view in the wrong way: The emphasis should be on the effects of properties of alleles, not on allelic properties of organisms (like the property of having an *A* allele) and the accounting ought to be done in terms of allele copies. Second, although we argued above that the strategy of splitting populations was at odds with the character of evolutionary theory, it is worth noting that the same strategy will be available in the heterozygote superiority case.

Consider the following division of the original population: let P_1 be the collection of all those allele copies which occur next to an S allele, and let P_2 consist of all those allele copies which occur next to an A allele. Then the property of being A (or of directing the production of normal hemoglobin) has a positive effect on the production of copies in the next generation in P_1, and conversely in P_2. In this way, we are able to partition the population and to achieve a Dawkinsian redescription that meets Sober's principle (A) – just in the way that we might try to do so if we wanted to satisfy (A) in understanding the operation of selection on melanism in a Cheshire wood or on fighting strategies in a population containing a mixture of hawks and doves.

Objection: The "populations" just defined are highly unnatural, and this can be seen once we recognize that, in some cases, allele copies in the same organisms (the heterozygotes) belong to different "populations." Reply: So what? From the allele's point of view, the copy next door is just a critical part of the environment. The populations P_1 and P_2 simply pick out the alleles that share the same environment. There would be an analogous partition of a population of competing organisms which occurred locally in pairs such that some organisms played dove and some hawk. (Here, mixed pairs would correspond to heterozygotes.)

So the genic picture survives an important initial challenge. The moral of our story so far is that the picture must be applied consistently. Just as paradoxical conclusions will result if one offers a partial translation of geometry into arithmetic, it is possible to generate perplexities by failing to recognise that the Dawkinsian *Weltanschauung* leads to new conceptions of environment and of population. We now turn to a different worry, the objection that genes are not "visible" to selection.

II EPIGENESIS AND VISIBILITY

In a lucid discussion of Dawkins's early views, Gould claims to find a "fatal flaw" in the genic approach to selection. According to Gould, Dawkins is unable to give genes "direct visibility to natural selection."[7] Bodies must play intermediary roles in the process of selection, and,

36

since the properties of genes do not map in one-one fashion onto the properties of bodies, we cannot attribute selective advantages to individual alleles. We believe that Gould's concerns raise two important kinds of issues for the genic picture: (i) Can Dawkins sensibly talk of the effect of an individual allele on its expected copying frequency? (ii) Can Dawkins meet the charge that it is the phenotype that makes the difference to the copying of the underlying alleles, so that, whatever the causal basis of an advantageous trait, the associated allele copies will have enhanced chances of being replicated? We shall take up these questions in order.

Do Alleles Have Effects? Dawkins and Gould agree on the facts of embryology which subvert the simple Mendelian association of one gene with one character. But the salience of these facts to the debate is up for grabs. Dawkins regards Gould as conflating the demands of embryology with the demands of the theory of evolution. While genes' effects blend in embryological development, and while they have phenotypic effects only in concert with their gene-mates, genes "do not blend as they replicate and recombine down the generations. It is this that matters for the geneticist, and it is also this that matters for the student of units of selection" (Dawkins 1982, p. 117).

Is Dawkins right? Chapter 2 of his 1982 is an explicit defence of the meaningfulness of talk of "genes for" indefinitely complex morphological and behavioural traits. In this, we believe, Dawkins is faithful to the practice of classical geneticists. Consider the vast number of loci in *Drosophila melanogaster* which are labeled for eye-color traits – white, eosin, vermilion, raspberry, and so forth. Nobody who subscribes to this practice of labelling believes that a pair of appropriately chosen stretches of DNA, cultured in splendid isolation, would produce a detached eye of the pertinent colour. Rather, the intent is to indicate the effect that certain changes at a locus would make against the background of the rest of the genome.

Dawkins's project here is important not just in conforming to traditions of nomenclature. Remember: Dawkins needs to show that we can sensibly speak of alleles having (environment-sensitive) effects, effects in virtue of which they are selected for or selected against. If we can talk of a gene for X, where X is a selectively important phenotypic characteristic, we can sensibly talk of the effect of an allele on its expected copying frequency, even if the effects are always indirect, via the characteristics of some vehicle.

What follows is a rather technical reconstruction of the relevant notion. The precision is needed to allow for the extreme environmental sensitivity of allelic causation. But the intuitive idea is simple: We can speak of genes for X if substitutions on a chromosome would lead, in the relevant environments, to a difference in the X-ishness of the phenotype.

Consider a species S and an arbitrary locus L in the genome of members of S. We want to give sense to the locution 'L is a locus affecting P' and derivatively to the phrase 'G is a gene for P^*' (where, typically, P will be a determinable and P^* a determinate form of P). Start by taking an *environment* for a locus to be an aggregate of DNA segments that would complement L to form the genome of a member of S together with a set of extra-organismic factors (those aspects of the world external to the organism which we would normally count as part of the organism's environment). Let a set of variants for L be any collection of DNA segments, none of which is debarred, on physico-chemical grounds, from occupying L. (This is obviously a very weak constraint, intended only to rule out those segments which are too long or which have peculiar physico-chemical properties.) Now, we say that L is a locus affecting P in S relative to an environment E and a set of variants V just in case there are segments s, s^*, and s^{**} in V such that the substitution of s^{**} for s^* in an organism having s and s^* at L would cause a difference in the form of P, against the background of E. In other words, given the environment E, organisms who are ss^* at L differ in the form of P from organisms who are ss^{**} at L and the cause of the difference is the presence of s^* rather than s^{**}. (A minor clarification: while s^* and s^{**} are distinct, we do not assume that they are both different from s.)

L is a locus affecting P in S just in case L is a locus affecting P in S relative to any standard environment and a feasible set of variants. Intuitively, the geneticist's practice of labelling loci focuses on the "typical" character of the complementary part of the genome in the species, the "usual" extra-organismic environment, and the variant DNA segments which have arisen in the past by mutation or which "are likely to arise" by mutation. Can these vague ideas about standard conditions be made more precise? We think so. Consider first the genomic part of the environment. There will be numerous alternative combinations of genes at the loci other than L present in the species S. Given most of these gene combinations, we expect modifications at L to produce modifications in the form of P. But there are likely to be some exceptions, cases in which the presence of a rare allele at another

38

locus or a rare combination of alleles produces a phenotypic effect that dominates any effect on P. We can either dismiss the exceptional cases as non-standard because they are infrequent or we can give a more refined analysis, proposing that each of the non-standard cases involves either (a) a rare allele at a locus L' or (b) a rare combination of alleles at loci L', L'' . . . such that that locus (a) or those loci jointly (b) affect some phenotypic trait Q that dominates P in the sense that there are modifications of Q which prevent the expression of any modification of P. As a concrete example, consider the fact that there are modifications at some loci in *Drosophila* which produce embryos that fail to develop heads; given such modifications elsewhere in the genome, alleles affecting eye color do not produce their standard effects!

We can approach standard extra-genomic environments in the same way. If L affects the form of P in organisms with a typical gene complement, except for those organisms which encounter certain rare combinations of external factors, then we may count those combinations as non-standard simply because of their infrequency. Alternatively, we may allow rare combinations of external factors to count provided that they do not produce some gross interference with the organism's development, and we can render the last notion more precise by taking non-standard environments to be those in which the population mean fitness of organisms in S would be reduced by some arbitrarily chosen factor (say, $1/2$).

Finally, the feasible variants are those which actually occur at L in members of S, together with those which have occurred at L in past members of S and those which are easily attainable from segments that actually occur at L in members of S by means of insertion, deletion, substitution, or transposition. Here the criteria for ease of attainment are given by the details of molecular biology. If an allele is prevalent at L in S, then modifications at sites where the molecular structure favors insertions, deletions, substitutions, or transpositions (so-called hot spots) should count as easily attainable even if some of these modifications do not actually occur.

Obviously, these concepts of "standard conditions" could be articulated in more detail, and we believe that it is possible to generate a variety of explications, agreeing on the core of central cases but adjusting the boundaries of the concepts in different ways. If we now assess the labelling practices of geneticists, we expect to find that virtually all of their claims about loci affecting a phenotypic trait are sanctioned by all of the explications. Thus, the challenge that there is no way to honour the facts of epigenesis while speaking of loci that affect certain traits would be turned back.

Once we have come this far, it is easy to take the final step. An allele A at a locus L in a species S is for the trait $P*$ (assumed to be a determinate form of the determinable characteristic P) relative to a local allele B and an environment E just in case (a) L affects the form of P in S, (b) E is a standard environment, and (c) in E organisms that are AB have phenotype $P*$. The relativisation to a local allele is necessary, of course, because, when we focus on a target allele rather than a locus, we have to extend the notion of the environment – as we saw in the last section, corresponding alleles are potentially important parts of one another's environments. If we say that A is for $P*$ (period), we are claiming that A is for $P*$ relative to standard environments and common local alleles or that A is for $P*$ relative to standard environments and itself.

Now, let us return to Dawkins and to the apparently outré claim that we can talk about genes for reading. Reading is an extraordinarily complex behaviour pattern and surely no adaptation. Further, many genes must be present and the extra-organismic environment must be right for a human being to be able to acquire the ability to read. Dyslexia might result from the substitution of an unusual mutant allele at one of the loci, however. Given our account, it will be correct to say that the mutant allele is a gene for dyslexia and also that the more typical alleles at the locus are alleles for reading. Moreover, if the locus also affects some other (determinable) trait, say, the capacity to factor numbers into primes, then it may turn out that the mutant allele is also an allele for rapid factorisation skill and that the typical allele is an allele for factorisation disability. To say that A is an allele for $P*$ does not preclude saying that A is an allele for $Q*$, nor does it commit us to supposing that the phenotypic properties in question are either both skills or both disabilities. Finally, because substitutions at many loci may produce (possibly different types of) dyslexia, there may be many genes for dyslexia and many genes for reading. Our reconstruction of the geneticists' idiom, the idiom which Dawkins wants to use, is innocent of any Mendelian theses about one-one mappings between genes and phenotypic traits.

Visibility. So we can defend Dawkins's thesis that alleles have properties that influence their chances of leaving copies in later generations by suggesting that, in concert with their environments (including their genetic environments), those alleles cause the presence of certain properties in vehicles (such as organisms) and that the properties of the vehicles are causally relevant to the spreading of copies of the alleles. But our answer to question (i) leads naturally to concerns about question (ii). Granting that an allele is for a phenotypic trait P^* and that the presence of P^* rather than alternative forms of the determinable trait P enhances the chances that an organism will survive and reproduce and thus transmit copies of the underlying allele, is it not P^* and its competition which are directly involved in the selection process? What selection "sees" are the phenotypic properties. When this vague, but suggestive, line of thought has been made precise, we think that there is an adequate Dawkinsian reply to it.

The idea that selection acts directly on phenotypes, expressed in metaphorical terms by Gould (and earlier by Ernst Mayr), has been explored in an interesting essay by Robert Brandon.[8] Brandon proposes that phenotypic traits screen off genotypic traits (in the sense of Wesley Salmon[9]):

$$\Pr(O_n/G\,\&\,P) = \Pr(O_n/P) \neq \Pr(O_n/G)$$

where $\Pr(O_n/G\&P)$ is the probability that an organism will produce n offspring given that it has both a phenotypic trait and the usual genetic basis for that trait, $\Pr(O_n/P)$ is the probability that an organism will produce n offspring given that it has the phenotypic trait, and $\Pr(O_n/G)$ is the probability that it will produce n offspring given that it has the usual genetic basis. So fitness seems to vary more directly with the phenotype and less directly with the underlying genotype.

Why is this? The root idea is that the successful phenotype may occur in the presence of the wrong allele as a result of judicious tampering, and, conversely, the typical effect of a "good" allele may be subverted. If we treat moth larvae with appropriate injections, we can produce pseudomelanics that have the allele which normally gives rise to the speckled form and we can produce moths, foiled melanics, that carry the allele for melanin in which the developmental pathway to the emergence of black wings is blocked. The pseudomelanics will enjoy enhanced reproductive success in polluted woods and the foiled melanics will be at a disadvantage. Recognizing this type of possibility, Brandon concludes that selection acts at the level of the phenotype.[10]

Once again, there is no dispute about the facts. But our earlier discussion of epigenesis should reveal how genic selectionists will want to tell a different story. The interfering conditions that affect the phenotype of the vehicle are understood as parts of the allelic environment. In effect, Brandon, Gould, and Mayr contend that, in a polluted wood, there is selection for being dark coloured rather than for the allelic property of directing the production of melanin, because it would be possible to have the reproductive advantage associated with the phenotype without having the allele (and conversely it would be possible to lack the advantage while possessing the allele). Champions of the gene's-eye view will maintain that tampering with the phenotype reverses the typical effect of an allele by changing the environment. For these cases involve modification of the allelic environment and give rise to new selection processes in which allelic properties currently in favor prove detrimental. The fact that selection goes differently in the two environments is no more relevant than the fact that selection for melanic colouration may go differently in Cheshire and in Doreset.

If we do not relativize to a fixed environment, then Brandon's claims about screening off will not generally be true.[11] We suppose that Brandon intends to relativise to a fixed environment. But now he has effectively begged the question against the genic selectionist by deploying the orthodox conception of environment. Genic selectionists will also want to relativise to the environment, but they should resist the orthodox conception of it. On their view, the probability relations derived by Brandon involve an illicit averaging over environments (see fn. 12). Instead, genic selectionists should propose that the probability of an allele's leaving n copies of itself should be understood relative to the total allelic environment, and that the specification of the total environment ensures that there is no screening off of allelic properties by phenotypic properties. The probability of producing n copies of the allele for melanin in a total allelic environment is invariant under conditionalisation on phenotype.

Here too the moral of our story is that Dawkinspeak must be undertaken consistently. Mixing orthodox concepts of the environment with ideas about genic selection is a recipe for trouble, but we have tried to show how the genic approach can be thoroughly articulated so as to meet major objections. But what is the point of doing so? We shall close with a survey of some advantages and potential drawbacks.

III GENES AND GENERALITY

Relatively little fossicking is needed to uncover an extended defence of the view that gene selectionism offers a more general and unified picture of selective processes than can be had from its alternatives. Phenomena anomalous for the orthodox story of evolution by individual selection fall naturally into place from Dawkins's viewpoint. He offers a revision of the "central theorem" of Darwinism. Instead of expecting individuals to act in their best interests, we should expect an animal's behaviour "to maximize the survival of genes 'for' that behavior, whether or not those genes happen to be in the body of that particular animal performing it" (Dawkins 1982, p. 223).

The cases that Dawkins uses to illustrate the superiority of his own approach are a somewhat motley collection. They seem to fall into two general categories. First are outlaw and quasi-outlaw examples. Here there is competition among genes which cannot be translated into talk of vehicle fitness because the competition is among co-builders of a single vehicle. The second group comprises "extended phenotype" cases, instances in which a gene (or combination of genes) has selectively relevant phenotypic consequences which are not traits of the vehicle that it has helped build. Again the replication potential of the gene cannot be translated into talk of the adaptedness of its vehicle.

We shall begin with outlaws and quasi-outlaws. From the perspective of the orthodox story of individual selection, "replicators at different loci within the same body can be expected to 'cooperate'." The allele surviving at any given locus tends to be the one best (subject to all the constraints) for the whole genome. By and large this is a reasonable assumption. Whereas individual outlaw organisms are perfectly possible in groups and subvert the chances for groups to act as vehicles, outlaw genes seem problematic. Replication of any gene in the genome requires the organisms to survive and reproduce, so genes share a substantial common interest. This is true of asexual reproduction, and, granting the fairness of meiosis, of sexual reproduction too.

But there is the rub. Outlaw genes are genes which subvert meiosis to give them a better than even chance of making it to the gamete, typically by sabotaging their corresponding allele (Dawkins 1982, p. 136). Such genes are *segregation distorters* or *meiotic drive* genes. Usually, they are enemies not only of their alleles but of other parts

43

of the genome, because they reduce the individual fitness of the organism they inhabit. Segregation distorters thrive, when they do, because
they exercise their phenotypic power to beat the meiotic lottery. Selection for such genes cannot be selection for traits that make organisms
more likely to survive and reproduce. They provide uncontroversial
cases of selective processes in which the individualistic story cannot
be told.

There are also related examples. Altruistic genes can be outlaw-like,
discriminating against their genome mates in favour of the inhabitants
of other vehicles, vehicles that contain copies of themselves. Start with
a hypothetical case, the so-called green beard effect. Consider a gene
Q with two phenotypic effects. Q causes its vehicle to grow a green
beard and to behave altruistically towards green-bearded conspecifics.
Q's replication prospects thus improve, but the particular vehicle that
Q helped build does not have its prospects for survival and reproduction enhanced. Is Q an outlaw not just with respect to the vehicle but
with respect to the vehicle builders? Will there be selection for alleles
that suppress Q's effect? How the selection process goes will depend
on the probability that Q's co-builders are beneficiaries as well. If Q is
reliably associated with other gene kinds, those kinds will reap a net
benefit from Q's outlawry.

So altruistic genes are sometimes outlaws. Whether coalitions of
other genes act to suppress them depends on the degree to which they
benefit only themselves. Let us now move from a hypothetical example
to the parade case.

Classical fitness, an organism's propensity to leave descendants in
the next generation, seems a relatively straightforward notion. Once it
was recognized that Darwinian processes do not necessarily favour
organisms with high classical fitness, because classical fitness ignores
indirect effects of costs and benefits to relatives, a variety of alternative measures entered the literature. The simplest of these would be
to add to the classical fitness of an organism contributions from the
classical fitness of relatives (weighted in each case by the coefficient of
relatedness). Although accounting of this sort is prevalent, Dawkins
(rightly) regards it as just wrong, for it involves double book-keeping
and, in consequence, there is no guarantee that populations will move
to local maxima of the defined quantity. This measure and measures
akin to it, however, are prompted by Hamilton's rigorous development
of the theory of inclusive fitness (in which it is shown that populations
will tend towards local maxima of inclusive fitness).[12] In the mis-

understanding and misformulation of Hamilton's ideas, Dawkins sees an important moral.

Hamilton, he suggests, appreciated the gene selectionist insight that natural selection will favour "organs and behavior that cause the individual's genes to be passed on, whether or not the individual is an ancestor" (1982, p. 185). But Hamilton's own complex (and much misunderstood) notion of inclusive fitness was, for all its theoretical importance, a dodge, a "brilliant last-ditch rescue attempt to save the individual organism as the level at which we think about natural selection" (1982, p. 187). More concretely, Dawkins is urging two claims: first, that the uses of the concept of inclusive fitness in practice are difficult, so that scientists often make mistakes; second, that such uses are conceptually misleading. The first point is defended by identifying examples from the literature in which good researchers have made errors, errors which become obvious once we adopt the gene selectionist perspective. Moreover, even when the inclusive fitness calculations make the right predictions, they often seem to mystify the selective process involved (thus buttressing Dawkins's second thesis). Even those who are not convinced of the virtues of gene selectionism should admit that it is very hard to see the reproductive output of an organism's relatives as a property of that organism.

Let us now turn to the other family of examples, the "extended phenotype" cases. Dawkins gives three sorts of "extended" phenotypic effects: effects of genes – indeed key weapons in the competitive struggle to replicate – which are not traits of the vehicle the genes inhabit. The examples are of artefacts, of parasitic effects on host bodies and behaviours, and of "manipulation" (the subversion of an organism's normal patterns of behaviour by the genes of another organism via the manipulated organism's nervous system).

Among many vivid, even haunting, examples of parasitic behaviour, Dawkins describes cases in which parasites synthesise special hormones with the consequence that their hosts take on phenotypic traits that decrease their own prospects for reproduction but enhance those of the parasites (see, for a striking instance, 1982, p. 215). There are equally forceful cases of manipulation: cuckoo fledglings subverting their host's parental programme, parasitic queens taking over a hive and having its members work for her. Dawkins suggests that the traits in question should be viewed as adaptations – properties for which selection has occurred – even though they cannot be seen as adaptations of the individuals whose reproductive success they promote, for

45

those individuals do not possess the relevant traits. Instead, we are to think in terms of selectively advantageous characteristics of alleles which orchestrate the behaviour of several different vehicles, some of which do not include them.

At this point there is an obvious objection. Can we not understand the selective processes that are at work by focusing not on the traits that are external to the vehicle that carries the genes, but on the behaviour that the vehicle performs which brings those traits about? Consider a spider's web. Dawkins wants to talk of a gene for a web. A web, of course, is not a characteristic of a spider. Apparently, however, we could talk of a gene for web building. Web building is a trait of spiders, and, if we choose to redescribe the phenomena in these terms, the extended phenotype is brought closer to home. We now have a trait of the vehicle in which the genes reside, and we can tell an orthodox story about natural selection for this trait.

It would be tempting to reply to this objection by stressing that the selective force acts through the artefact. The causal chain from the gene to the web is complex and indirect; the behaviour is only a part of it. Only one element of the chain is distinguished, the endpoint, the web itself, and that is because, independently of what has gone on earlier, provided that the web is in place, the enhancement of the replication chances of the underlying allele will ensue. But this reply is exactly parallel to the Mayr–Gould–Brandon argument discussed in the last section, and it should be rejected for exactly parallel reasons.

The correct response, we believe, is to take Dawkins at his word when he insists on the possibility of a number of different ways of looking at the same selective processes. Dawkins's two main treatments of natural selection, 1976 and 1982, offer distinct versions of the thesis of genic selectionism. In the earlier discussion (and occasionally in the later) the thesis is that, for any selection process, there is a uniquely correct representation of that process, a representation which captures the causal structure of the process, and this representation attributes causal efficacy to genic properties. In 1982, especially in Chapters 1 and 13, Dawkins proposes a weaker version of the thesis, to the effect that there are often alternative, equally adequate representations of selection processes and that, for any selection process, there is a maximally adequate representation which attributes causal efficacy to genic properties. We shall call the strong (early) version *monist genic selectionism* and the weak (later) version *pluralist genic selectionism*. We believe that the monist version is faulty but that the pluralist thesis is defensible.

In presenting the "extended phenotype" cases, Dawkins is offering an alternative representation of processes that individualists can redescribe in their own preferred terms by adopting the strategy illustrated in our discussion of spider webs. Instead of talking of genes for webs and their selective advantages, it is possible to discuss the case in terms of the benefits that accrue to spiders who have a disposition to engage in web building. There is no privileged way to segment the causal chain and isolate the (really) real causal story. As we noted two paragraphs back, the analog of the Mayr–Gould–Brandon argument for the priority of those properties which are most directly connected with survival and reproduction – here the webs themselves – is fallacious. Equally, it is fallacious to insist that the causal story must be told by focusing on traits of individuals which contribute to the reproductive success of those individuals. We are left with the general thesis of pluralism: there are alternative, maximally adequate representations of the causal structure of the selection process. Add to this Dawkins's claim that one can always find a way to achieve a representation in terms of the causal efficacy of genic properties, and we have pluralist genic selectionism.

Pluralism of the kind we espouse has affinities with some traditional views in the philosophy of science. Specifically, our approach is instrumentalist, not of course in denying the existence of entities like genes, but in opposing the idea that natural selection is a force that acts on some determinate target, such as the genotype or the phenotype. Monists err, we believe, in claiming that selection processes must be described in a particular way, and their error involves them in positing entities, "targets of selection," that do not exist.

Another way to understand our pluralism is to connect it with conventionalist approaches to space–time theories. Just as conventionalists have insisted that there are alternative accounts of the phenomena which meet all our methodological desiderata, so too we maintain that selection processes can usually be treated, equally adequately, from more than one point of view. The virtue of the genic point of view, on the pluralist account, is not that it alone gets the causal structure right but that it is always available.

What is the rival position? Well, it cannot be the thesis that the only adequate representations are those in terms of individual traits which promote the reproductive success of their bearers, because there are instances in which no such representation is available (outlaws) and instances in which the representation is (at best) heuristically misleading (quasi-outlaws, altruism). The sensible rival position is that there is a hierarchy of selection processes: some cases are aptly represented in terms of genic selection, some in terms of individual selection, some in terms of group selection, and some (maybe) in terms of species selection. Hierarchical monism claims that, for any selection process, there is a unique level of the hierarchy such that only representations that depict selection as acting at that level are maximally adequate. (Intuitively, representations that see selection as acting at other levels get the causal structure wrong.) Hierarchical monism differs from pluralist genic selectionism in an interesting way: whereas the pluralist insists that, for any process, there are many adequate representations, one of which will always be a genic representation, the hierarchical monist maintains that for each process there is just one kind of adequate representation, but that processes are diverse in the kinds of representation they demand.[13]

Just as the simple orthodoxy of individualism is ambushed by outlaws and their kin, so, too, hierarchical monism is entangled in spider webs. In the "extended phenotype" cases, Dawkins shows that there are genic representations of selection processes which can be no more adequately illuminated from alternative perspectives. Since we believe that there is no compelling reason to deny the legitimacy of the individualist redescription in terms of web-building behaviour (or dispositions to such behaviour), we conclude that Dawkins should be taken at face value: just as we can adopt different perspectives on a Necker cube, so too we can look at the workings of selection in different ways (1982, Chapter 1).

In previous sections, we have tried to show how genic representations are available in cases that have previously been viewed as troublesome. To complete the defence of genic selectionism, we would need to extend our survey of problematic examples. But the general strategy should be evident. Faced with processes that others see in terms of group selection or species selection, genic selectionists will first try to achieve an individualist representation and then apply the ideas we have developed from Dawkins to make the translation to genic terms.

Pluralist genic selectionists recommend that practicing biologists take advantage of the full range of strategies for representing the workings of selection. The chief merit of Dawkinspeak is its generality. Whereas the individualist perspective may sometimes break down, the gene's-eye view is apparently always available. Moreover, as illustrated by the treatment of inclusive fitness, adopting it may sometimes help us to avoid errors and confusions. Thinking of selection in terms of the devices, sometimes highly indirect, through which genes level themselves into future generations may also suggest new approaches to familiar problems.

But are there drawbacks? Yes. The principal purpose of the early sections of this chapter was to extend some of the ideas of genic selectionism to respond to concerns that are deep and important. Without an adequate rethinking of the concepts of population and of environment, genic representations will fail to capture processes that involve genic interactions or epigenetic constraints. Genic selectionism can easily slide into naive adaptationism as one comes to credit the individual alleles with powers that enable them to operate independently of one another. The move from the "genes for *P*" locution to the claim that selection can fashion *P* independently of other traits of the organism is perennially tempting.[14] But, in our version, genic representations must be constructed in full recognition of the possibilities for constraints in gene-environment co-evolution. The dangers of genic selectionism, illustrated in some of Dawkins's own writings, are that the commitment to the complexity of the allelic environment is forgotten in practice. In defending the genic approach against important objections, we have been trying to make this commitment explicit, and thus to exhibit both the potential and the demands of correct Dawkinspeak. The return of the gene should not mean the exile of the organism.[15]

NOTES

1. The claim is made in *The Selfish Gene* (New York: Oxford, 1976); and, in a somewhat modified form, in *The Extended Phenotype* (San Francisco: Freeman, 1982). We shall discuss the difference between the two versions in the final section of this chapter, and our reconstruction will be primarily concerned with the later version of Dawkins's thesis. To forestall any possible confusion, our reconstruction of Dawkins's position does not commit us to the provocative claims about altruism and selfishness on which many early critics of Dawkins (1996) fastened.
2. See Sober, 1984, Chs. 7–9, especially pp. 302–14.
3. For details, see Maynard Smith (1982); and, for a capsule presentation, Kitcher (1985), pp. 88–97.
4. The *locus classicus* for discussion of this example is Kettlewell (1973).

5. In the spirit of Sober's original argument, one might press further. Genic selectionists contend that an A allele can find itself in two different environments, one in which the effect of directing the formation of a normal globin chain is positive and one in which that effect is negative. Should we not be alarmed by the fact that the distribution of environments in which alleles are selected is itself a function of the frequency of the alleles whose selection we are following? No. The phenomenon is thoroughly familiar from studies of behavioural interactions – in the hawk-dove case we treat the frequency of hawks both as the variable we are tracking and as a facet of the environment in which selection occurs. Maynard Smith makes the parallel fully explicit in his 1987 pp. 119–31, especially pp. 125–6.

6. Moreover, we can explicitly recognise the co-evolution of alleles with allelic environments. A fully detailed general approach to population genetics from the Dawkinsian point of view will involve equations that represent the functional dependence of the distribution of environments on the frequency of alleles, and equations that represent the fitnesses of individual alleles in different environments. In fact, this is just another way of looking at the standard population genetics equations. Instead of thinking of W_{AA} as the expected contribution to survival and reproduction of (an organism with) an allelic pair, we think of it as the expected contribution of copies of itself of the allele A in environment A. We now see W_{AS} as the expected contribution of A in environment S and also as the expected contribution of S in environment A. The frequencies p, q are not only the frequencies of the alleles, but also the frequencies with which certain environments occur. The standard definitions of the overall (net) fitnesses of the alleles are obtained by weighting the fitnesses in the different environments by the frequencies with which the environments occur.

   Lewontin has suggested to us that problems may arise with this scheme of interpretation if the population should suddenly start to reproduce asexually. But this hypothetical change could be handled from the genic point of view by recognising an alteration of the co-evolutionary process between alleles and their environments: whereas certain alleles used to have descendants that would encounter a variety of environments, their descendants are now found only in one allelic environment. Once the algebra has been formulated, it is relatively straightforward to extend the reinterpretation to this case.

7. Gould (1980e), p. 90. There is a valuable discussion of Gould's claims in Sober (1984), p. 227ff.

8. Gould 1980e; Mayr, 1963, p. 184; and Brandon, 1984, pp. 133–41.

9. Brandon refers to Salmon (1971). It is now widely agreed that statistical relevance misses some distinctions which are important in explicating causal relevance. See, for example, Cartwright (1979), Sober (1984), and Salmon (1984).

10. Unless the treatments are repeated in each generation, the presence of the genetic basis for melanic colouration will be correlated with an increased frequency of grandoffspring, or of great-grandoffspring, or of descendants in some further generation. Thus, analogs of Brandon's probabilistic relations

will hold only if the progeny of foiled melanics are treated so as to become foiled melanics, and the progeny of pseudomelanics are treated so as to become pseudomelanics. This point reinforces the claims about the relativisation to the environment that we make below. Brandon has suggested to us in correspondence that now his preferred strategy for tackling issues of the units of selection would be to formulate a principle for identifying genuine environments.

11. Intuitively, this will be because Brandon's identities depend on there being no correlation between O_n and G in any environment, except through the property P. Thus, ironically, the screening-off relations only obtain under the assumptions of simple bean-bag genetics! Sober seems to appreciate this point in a cryptic footnote in Sober (1984), pp. 229–30.

To see how it applies in detail, imagine that we have more than one environment and that the reproductive advantages of melanic colouration differ in the different environments. Specifically, suppose that E_1 contains m_1 organisms that have P (melanic colouration) and G (the normal genetic basis of melanic colouration), that E_2 contains m_2 organisms that have P and G, and that the probabilities $\Pr(O_n/G\&P\&E_1)$ and $\Pr(O_n/G\&P\&E_2)$ are different. Then, if we do not relativise to environments, we shall compute $\Pr(O_n/G\&P)$ as a weighted average of the probabilities relative to the two environments.

$$\begin{aligned}
\Pr(O_n/G\&P) &= \Pr(E_1/G\&P)\cdot\Pr(O_n/G\&P\&E_1) \\
&\quad + \Pr(E_2/G\&P)\cdot\Pr(O_n/G\&P\&E_2) \\
&= m_1/(m_1+m_2)\cdot\Pr(O_n/G\&P\&E_1) \\
&\quad + m_2/(m_1+m_2)\cdot\Pr(O_n/G\&P\&E_2)
\end{aligned}$$

Now, suppose that tampering occurs in E_2 so that there are m_3 pseudomelanics in E_2. We can write $\Pr(O_n/P)$ as a weighted average of the probabilities relative to the two environments.

$$\Pr(O_n/P) = \Pr(E_1/P)\cdot\Pr(O_n/P\&E_1) + \Pr(E_2/P)\cdot\Pr(O_n/P\&E_2).$$

By the argument that Brandon uses to motivate his claims about screening off, we can take $\Pr(O_n/G\&P\&E_i) = \Pr(O_n/P\&E_i)$ for $i = 1, 2$. However, $\Pr(E_1/P) = m_1/(m_1 + m_2 + m_3)$ and $\Pr(E_2/P) = (m_2 + m_3)/(m_1 + m_2 + m_3)$, so that $\Pr(E_i/P) \neq \Pr(E_i/G\&P)$. Thus, $\Pr(O_n/G\&P) \neq \Pr(O_n/P)$, and the claim about screening off fails.

Notice that, if environments are lumped in this way, then it will only be under fortuitous circumstances that the tampering makes the probabilistic relations come out as Brandon claims. Pseudo-melanics would have to be added in both environments so that the weights remain exactly the same.

12. For Hamilton's original demonstration, see Hamilton (1971). For a brief presentation of Hamilton's ideas, see Kitcher (1985), pp. 77–87; and for penetrating diagnoses of misunderstandings, see Grafen (1982); and Michod (1984).

13. In defending pluralism, we are very close to the views expressed by Maynard

Smith (1987). Indeed, we would like to think that Maynard Smith's article and the present essay complement one another in a number of respects. In particular, as Maynard Smith explicitly notes, "recommending a plurality of models of the same process" contrasts with the view (defended by Gould and by Sober) of "emphasizing a plurality of processes." Gould's views are clearly expressed in Gould (1980d); and Sober's ideas are presented in Sober (1984), Chapter 9.

14. At least one of us believes that the claims of the present chapter are perfectly compatible with the critique of adaptationism developed in Gould and Lewontin (1978). For discussion of the difficulties with adaptationism, see Kitcher (1985), Ch. 7; and Kitcher (1987).

15. As, we believe, Dawkins himself appreciates. See the last chapter of 1982, especially his reaction to the claim that "Richard Dawkins has rediscovered the organism" (p. 251).

16. This was co-authored with Philip Kitcher. We are equally responsible for this chapter which was written when we discovered that we were writing it independently. We would like to thank those who have offered helpful suggestions to one or both of us, particularly Patrick Bateson, Robert Brandon, Peter Godfrey-Smith, David Hull, Richard Lewontin, Lisa Lloyd, Philip Pettit, David Scheel, and Elliott Sober.

3

The Extended Replicator

I INTRODUCTION

Our purpose in this chapter is to evaluate a conception of evolution in general, and the units of selection in particular, that have been articulated by a group that we shall refer to as Developmental Systems Theorists.[1] So we first outline their distinctive "take" on evolution and the units of selection, and contrast it with three other perspectives. We then compare and contrast our views of genes and replicators with that of Developmental Systems Theory. In resisting their view that genes play no distinctive informational role in inheritance, it becomes clear that though the genes play a very special role in development, they are not alone in playing this role. We argue that despite its insights Developmental Systems Theory has serious problems. Moreover, its insights can be captured by a less radical take on the units of selection problem. We think that Dawkins (1982) and Hull (1981; 1988) were right to distinguish between replication and interaction, but we think they underestimate the range of biological replication. Finally, we speculate on that extended range, and suggest that there are good reasons for thinking that Bateson's famous reduction of the replicator is no reductio at all.

I.1   Four Evolutionary Mindsets

We begin by offering a low-resolution picture of the logical geography. We see four very general ways of characterising evolution.

1. The Received View takes evolution to be the result of competition between individual organisms varying in fitness by virtue of

heritable characteristics. Opinions within this camp differ on the relative importance of selection and drift, and on how to link historical accident with phylogenetic and developmental constraints. But evolution shapes populations by acting on individual organisms in virtue of their traits.

2. The Gene's Eye sees evolution as the result of competition between genes that differ primarily in their capacities to affect interactors. These interactors' differential reproduction results in differences in the depth and bushiness of germline lineages.

3. The Extended Replicator hijacks these conceptual tools without any commitment to the idea that genes are the sole, or even the predominant, replicator. The Gene's Eye fades into this conception as we recognise more and more non-genetic replicators. In this debate Dawkins is an important but equivocal figure. We think this chapter is a development of one of the threads of his work, for it has never been part of his official definition of "replicator" that only genes are replicators. The other category he recognises are memes, elements of culture, especially human culture. So one picture that descends from his work is that change results from two rather distinct evolutionary processes. Biological evolution works out the fate of competing lineages of genes. Social evolution does the same for competing lineages of memes. Instead, we think a single evolutionary process determines the fate of lineages of replicators of many kinds, by virtue of the differential success of their associated interactors and extended phenotypic effects. This concept is certainly consistent with Dawkins's basic conceptual structure, but is discordant with at least some aspects of his actual practice, which has emphasised the gene to the exclusion of other replicators. (Dawkins 1976; 1982; 1989.)

4. Finally, there is the Developmental Systems conception of evolution. This conception is hard to characterise precisely, but the following elements seem diagnostic: (i) In the cycle from one developmental system to its successor, no element of the developmental matrix plays any privileged causal role; (ii) There is no theoretically significant distinction between internal and external factors. All are necessary, and though they are necessary in different ways, none of these different ways are special. Genes are but one important element of the developmental matrix. For some particular purposes, a focus on genes is indeed appropriate, but they have no general or overarching privilege; (iii) Defenders of this family of conceptions

are sceptical about the idea of the transmission of information through inheritance mechanisms, and the associated metaphor of the genetic program; and (iv) For them, the developmental system as a whole is the unit of selection. It is those that are rebuilt from cycle to cycle. Evolution, on this view, is the result of competition between lineages of developmental systems. Their view is radical in part because they conceive of the developmental matrix, the resources that are reduplicated from developmental cycle to developmental cycle, very broadly, including elements that are on most conceptions part of the environment.

Some versions of Developmental Systems Theory – for example, Griffiths and Gray (1994) – take the replicator to be the set of processes through which the developmental system is built, rather than the system itself. We do not think this distinction important in this context, though in some contexts it is. For example, in defending the idea of a gene's extended phenotype, Dawkins is concerned to emphasise the fact that adaptive phenotypic effects do not necessarily come bundled into discrete organisms. Here, the emphasis on process rather than object makes a point: the adaptation is the process through which, say, a chick manipulates its parent's behaviour. But when the "object" is the developmental system as a whole – all the entities and their relations that go into constructing an organism – we do not see that this is a distinction that makes a difference.

I.2   Common Ground

Three important ideas about evolution are common ground between our view of evolution and that of the Developmental Systems Theorists. Everyone agrees that the genome of a developing organism is not sufficient for the development of any of its characteristics. Even so, many evolutionists think the genome's causal role in the development is privileged. In commenting on an earlier version of this chapter, David Hull expressed a common intuition: with the exception of cultural transmission in a few groups of animals, genes are the only, or almost the only, cause of structure. Developmental Systems Theorists deny this. We think they are right to make the stronger claim that the gene plays no privileged causal role in the development of phenotype from genotype.

Second, one way to think of inheritance is to see it as a causal bridge from phenotype to phenotype. On the received picture, that bridge proceeds exclusively through the DNA. The only pathway of inheritance, the only bridge between the generations over which information flows from phenotype to phenotype, is the DNA in the gametes. In both our view and that of the Developmental Systems Theorists this picture is not just an idealisation; it is seriously mistaken.

Third, Developmental System's Theorists do not exile the gene in order to embrace the organism. The Gene's Eye, Extended Replicator and Developmental System's conceptions of evolution agree in not conceptualising evolution *only* as a history of organisms in competition. The scope of evolution is richer, and weirder, than lineages of organisms. So arguing about "Organismism," the idea that "the organism is the subject and object of evolution" (Lewontin 1985), is not the aim of this chapter.

I.3 Development and Replication

Many critics of the Gene's Eye have emphasised the complexity of development and the many: many relation between genotype and phenotype. Those worries are not our worries. We are not claiming development is so complex that genes are not replicators. Rather, our view is that there are other replicators beside genes. Nor is Developmental Systems Theory just a rehash of the many arguments from the failure of bean-bag genetics. It has a new take on the units of selection problem. Developmental Systems Theorists want to emphasise the interdependence of developmental and evolutionary theory. This interdependence is contested. For example, Dawkins (1982) has defended combining a particulate view of the evolutionary role of the gene with an interactionist view of their developmental role. Though genes interact in development, they have independent evolutionary trajectories. So an argument is needed to forge a link between development and interaction. We think there is such an argument. But it has not been very clear, so our best reconstruction of it is as follows.

Step 1. If we consider a lineage of organisms from developmental cycle to developmental cycle, we will see that on each cycle there are many repetitions of important elements of the previous cycle. Genes, cellular machinery of various kinds, morphological and physiological traits, behaviours, and social structures are reliably rebuilt, cycle by cycle. There are many constancies maintained through these lineages, constancies that permit selection to be cumulative.

Step 2. The developmental process through which each cycle repeats is fabulously complex with the effect of each element depending on the effects of many others. Even so, we think genes would be the beneficiaries of adaptation and hence the unit of replication, if genes controlled, directed or were the organising centre of development. For genes would then still have a privileged role in development. It would not be arbitrary to think of them as *the* replicators.

Step 3. The notion of genetic information and its relatives cannot be made good in a way that singles out the genes as having particular significance. Genes predict phenotypic characters only in the same sense that environmental factors predict them.

Step 4. No causal selection scheme picks out genes, either in the development of one phenotype in the lineage, nor in the recreation of that developmental factor in the next link in the lineage. In particular:

– directness does not single out genes. The processes through which genes are copied and through which they produce their phenotypic effects are highly indirect.
– causal asymmetry does not pick out the genes. It is true that every other factor in the cycle from link to link in the lineage depends on the capacity of (germline) genes to produce copies of themselves. But the capacity of germline genes to replicate equally depends on the reliable reproduction of a host of these other factors.
– causal responsibility for variance does not distinguish the role of the gene. Genes can be selected in virtue of their effects. For example, relativised to a normalised set of background conditions, the substitution of one gene for another may yield a boldly striped organism. So that gene is the "gene for bold stripes." But the same symmetrical comparison between variants, relativised to a normalised background, gives us incubation temperatures for traits, cellular chemicals for traits, and so on as causally responsible for variance.
– fidelity does not pick out the genes. Genes are not the only factors that re-appear with great reliability. Moreover, there is no fidelity threshold on replication.
– causal importance does not pick out the genes either. Gene-environment interactions are too messy and ill-behaved for us to say genes are the most important causes of some traits whereas the environment is the most important cause of others.

Step 5. They conclude that nothing singles out genes as special. Hence genes are not the replicators; whole developmental systems are.

In this paper, we propose getting off this inference bus at step 3 (Smith thinks one might be able to get off at step 4, too; see his 1994). The way we do this commits us to the view that genes plus a fair range of other elements in the developmental matrix are replicators. Against the Gene's Eye conception of evolution we argue that the genes play no unique informational role. We argue against Developmental Systems Theorists that the replicator/interactor distinction remains of value, that it captures all that their alternative captures and that the Developmental System's conception of evolution suffers from problems it has yet to resolve.

II DEVELOPMENTAL SYSTEMS THEORY

In this section we briefly characterise Developmental Systems Theory and our reservations about it. We think serious problems remain unresolved and that its important insights can be incorporated within a more traditional biological ontology. Moreover, that ontology continues to have independent motivation.

According to Developmental Systems Theory, genes are but one element of a developmental matrix which ranges from genes through proteins in the maternal cytoplasm to exposure to rank order in the local primate population. The whole set of elements, and in particular the relations between them, form a complex whole which is the unit of evolution and selection. Of course, in different systems different resources will play different roles. But there is no privileged gene-role; nor even a gene-role that some non-genes also play. The developmental system as a whole is the only replicator, and evolution is the differential success of lineages of replicators. The full range of developmental resources is the complex system replicated in development.

II.1 Holism

A prima facie problem for this conception of evolution is its apparent commitment to holism. Everything causally connects to everything else. Even so, we can understand something without understanding everything. So if developmental systems include everything causally relevant to development, they are too ill-defined to be a coherent active unit; they are too diffuse to be the objects of selection. Is Elvis Presley part of our developmental systems by virtue of his role in the development of our musical sensibilities? No biologically meaningful unit includes both Dickison and Presley. The intermeshing of causal connections, and transitivity of causation, will import the same problem for any other organism. As Griffiths and Gray (1994) realise, the defence of the developmental systems conception of evolution requires distinctions amongst the factors causally relevant in development. This distinction cannot depend on any *simple* measure of causal relevance; for there is none that will distinguish the role of Elvis from the role of early nutrition. But some causal influences are going to be part of the developmental system that made us, and others are not, on pain of all developmental systems reducing to one. That would be holism run amok.

II.2   Boundary and Other
Problems

The transitivity of causation is implicated in several problems, not just one. Elvis highlights the boundary problem. What events and processes go into the developmental system, given that not everything causally relevant can do so? A "cycle problem" brings into focus the lineage. Evolution, and most especially the evolution of adaptive complexity, depends on cumulative selection, and hence on lineages of similar organisms or organism surrogates. That problem is not difficult for any theory that privileges the organism, for the lineage is just a sequence of organisms related by descent. Nor is it a particularly serious problem for those who think of evolution as a struggle between lineages of genes. But one of the ideas that unites the Developmental Systems Theorists with others sceptical of the Received View is their refusal to define basic evolutionary processes by appeal to just one of the contingent products of evolution, the organism. The developmental system relies on the stable generation by generation reproduction of developmental resources. But developmental system generations are not to be identified with organism generations. So when do generations begin and end: do we count from bird to bird, egg to egg, or nesting hole to nesting hole? Cycles of developmental resources are not necessarily in sync. For birds that breed for more than one season, the nest cycle is shorter than the bird cycle. Other resources cycle slower than organism generations (e.g., the social group, many parasites' hosts). The cycle length of symbionts need not be the same. So on inspection, the developmental system replicating itself generation by generation seems perhaps a congerie of associated and perhaps co-evolving but still independent lineages; more a guerilla band than a regular battalion. There is a related problem with counting lineages. Is a ring of Mullerian mimics one developmental system or many? Is an ant-plant mutualism a single developmental system or several?

To the best of our knowledge, Griffiths and Gray (1993; 1994) make the only explicit attempts to solve these problems. We do not intend to offer a point by point discussion of their line of argument, for it is not our view that these problems are intractable. But we do think they are difficult, and aim to show that through a sketch of their treatment of the "boundary" problem. They argue that we must:

> distinguish . . . developmental outcomes which have evolutionary explanations from those that do not. The interactions that produce outcomes with evolutionary explanations are part of the developmental system. There is an evolutionary explanation of the fact that the authors . . . have a thumb on each hand. . . . The thumb is an evolved trait. But the fact that one of us has a scar on his left hand has no such explanation. The scar is an individual trait (we are referring of course to the trait of having a scar just thus and so, not the general ability to scar). The resources that produce the thumbs are part of the developmental system. Some of those that produced the scar . . . are not. (Griffiths and Gray 1994, p. 286)

Obviously, there is indeed a difference between the thumb and the scar. Though all of Griffiths's parents had thumbs, there is no reason to believe that they had scars on them. But phenotypic plasticity suggests that a reliance on parent/offspring similarity would draw the distinction between individual and evolved traits in the wrong place. The lyrebird's song is unique to each bird, for they are famous mimics, and pick up all manner of extraneous sounds, including those of humans, their animals and machines (Reilly 1991). Yet this does not seem to be an "individual" trait in the same sense that a scarred hand is. Moreover, there is a sense in which the scar has an evolutionary explanation: scarring events are "historically associated with" the human lineage. There is an evolutionary component of any individual scar construction. So we have our doubts about the robustness of their distinction.

III EXTENDING THE REPLICATOR

In this section, we argue that genes do play a very special role in development. But we also argue that genes are not the only developmental resources that play that special role. Some developmental factors do not just cause similarities between one developmental cycle and its successor. They have the form they do because they cause those similarities. They are the replicators. We do not distinguish them on the grounds that they are more important than other factors in the developmental process; rather, we distinguish them because they are adapted to play the role they do in development.

III.1 Farewell to Genetic Traits

Genes are not the primary cause of phenotypic traits. Perhaps, though, they represent genetic traits or carry information about them. Such seems to be the idea of those who have followed Mayr's lead in speaking of the genome as a program that directs development. But even the idea of genetically programmed traits is in trouble. For information is typically understood as dependence. A signal carries information about a source to the extent that characteristics of the signal co-vary with features of the source. So the genome carries information about a phenotype just so long as features of the genome co-vary with phenotypic characters. Phenotypic plasticity means that this co-variation is far from perfect. It improves, of course, if we hold the environment fixed. Aspects of the genome will co-vary well with traits *in an environment*. But as Johnston (1988) and Smith (1992) emphasise, a similar dependence holds between the environment and the developing phenotype. If we hold the genome fixed, there will be co-variation between environment and phenotype, and hence features of the environment will carry phenotypic information. The human genome carries information about, say, human skeletal structure. But so does the nutritional, biochemical and cellular environment of the foetus. The link between genome and developed system – holding environmental factors constant – is not unique. A network of necessary environmental factors – holding genetic background constant – correlates equally well with the developed system. So that network carries information about development in just the same sense that the genome does.

Moreover, the genome quite frequently correlates better not with the designed outcome of development but with various dysfunctional outcomes. Sterelny and Kitcher (1988) unpacked "a gene for X" by appeal to the gene's role in normal total environment. But Griffiths and Gray (1994) point out that this idea does not work. Most acorns rot, so acorn genomes correlate better with rotting than with growth. So if we are to talk of information, we should talk of the acorn genome carrying information about how to rot rather than grow, for it correlates *better* with rotting than growing.

We doubt that there is a quick fix for this problem. Of course, this example would collapse if there were a "gene for growing," a gene complex which had a better than 50 percent chance of growing. But it is most unlikely that there is any such gene complex. Of course, some gene complexes have a better chance of growing than others. But if information just is correlation, and all gene complexes correlate better with rotting than growing, then they carry the information about how to rot, not how to grow. No doubt we can take a more fine-grained view of oak environments, for there will be some circumstances, "microenvironments" (see Brandon 1990), in which some gene complexes will have a better than even chance of growing. But then it's the pair of gene complex and micro-environment which correlate with growing and hence which carries the information, not the gene complex itself.

Thus, any view which identifies the genome as a representation of the phenotype must be consistent with the interaction of the genome with the other factors of the developmental matrix. In virtue of this interaction, any element of the developmental matrix correlates with the developed phenotype if we hold constant the rest. No factor correlates with the developed phenotype unless we hold constant the rest. We can indeed speak of "the gene for red eyes." For relativised to a normalised set of background conditions, the substitution of that gene for a rival yields a red-eyed organism. But precisely the same comparison between variants, relativised to a normalised background, shows we can speak of incubation temperatures for traits, cellular chemicals for traits, and so on. For example, phenotypic plasticity in plants is common, and often manifested in a fine-tuned adaptive response to the specifics of a particular environment (Sultan 1987). Environmental variant, relativised to constant genetic background, predicts phenotypic variant.

The elements of the developmental matrix interact in ways that make it impossible to regard the environment as the mere trigger of a genetically caused process. The Buckeye butterfly (*Precis coenia*) shifts colour seasonally in ways that on first inspection fit the paradigm of a genetic process with an environmental trigger, and hence the conception of the genome as control centre. As the season advances, the colour pattern of emerging butterflies shifts from a tan ventral hindwing (WILDTYPE) in the Spring to a reddish hindwing (ROSA) in the autumn. This is not just an environmentally induced change, for it is easy to breed strains that express the reddish morph under all conditions. But nor is it a genetic subroutine with an environmental trigger, for butterflies of most genotypes can be induced to emerge as red morphs. There is no single environmental trigger: there are multiple ways to induce red shift in the population: both low temperatures or short daylengths. Most importantly of all, environmental influences interact: an inductive temperature under one daylength is not inductive under another and vice versa. So the causal structure of red shift in the population cannot be analysed as an almost universal genetic programme with redundancy built into the environmental triggering (Smith 1993b).

The genome is interdependent not just with the external environment. Parents contribute much more than genetic material to the developing organism. Gametes are not just packets of DNA. Even the sperm is more than a mere packet of DNA. Centrioles organise the axis of genetic segregation by migrating in cell division to opposite poles of the cell to serve as anchors for the filaments attaching to the chromosomes. They thus play an indispensable role in ensuring that the daughter cells get an equal number of chromosomes. Yet centrioles are transmitted parallel to the genetic material in the gametes. They are not built by gametic DNA (Glover, Gonzales, and Raff 1993).

For genes to become active in the construction of proteins, the coding sequences must be read out of these sequences to build the exons that code for proteins. This machinery is not just a causally necessary conduit – a more or less noisy channel – through which genetic information passes. Some sequences are ambiguous, so different exons can be constructed from the same transmitted sequences. Which exon is built, and hence which protein is coded for, depends on the cellular environment (Fogle 1990). Cellular machinery does not just play a role in allowing coding DNA to be read; it affects what is read.

III.2   Symmetry, Information and
Representation

Genes and other factors are thus interdependent in development. Consequently, genes do not correlate with developmental outcomes in any way that distinguishes them in the developmental system. Nevertheless, we think the genome does *represent* developmental outcomes. For representation depends not on correlation but function. The plans of a building are not the primary cause of a building or of its features. A plan may correlate better with graft, waste and overspending than with the actual traits of a building. Despite failures of correlation, and despite correlation without representation, plans represent buildings because that is their function. Some elements of the developmental matrix – the replicators – represent phenotypes in virtue of their functions.

The genome is one of the *designed mechanisms* in virtue of which phenotypes and genotypes duplicate themselves. Adaptation is seen in the proof-reading and repair mechanisms of the genes but not only there. For example, many arthropods are linked in obligatory symbiosis with micro-organisms on which they depend for growth, micro-organisms which are transmitted in the egg. This mechanism can be very precise. For example, in one species of aphid *Colophina arma* the micro-organisms are not transmitted in those eggs designed to be dwarf males or sterile female soldiers, for in these morphs no growth spurt is required (Morgan and Baumann 1994). This idea of a designed copying mechanism is the key to understanding the privileged role of the replicators in the total developmental matrix. Some parent-offspring similarities result from elements of the developmental matrix that have been selected to produce those similarities. Replicators exist because of those selection histories, and that distinguishes their role in development.

These functional differences are reflected in counter-factual differences within the developmental matrix. Thus both the replicators and the environmental factors correlate equally well with normal development. But there is an important asymmetry. In our view, the genes are not the only replicators, but let us for the moment focus on them. Consider a facultatively desert-adapted shrub; a shrub whose leaf structure and shape reduces water loss if grown in arid environments. Both aridity and the shrub's genome are necessary for that shrub's adaptive response to the environment. But that genome only exists because of the causal path (in that environment) from genome to desert-resistant shrub. By contrast, the aridity of the environment exists independently of the causal path (in that genetic environment) from arid conditions to desert-adapted shrub. One element of the developmental matrix exists only because of its role in the production of the plant lineage phenotype. That is why it has the function of producing that phenotype, and hence why it represents that phenotype.

A similar asymmetry lurks in the second problem. Acorn genomes correlate better with rotting than growing. But if all acorns rotted, there would be no development trajectory from acorn to worm food. In contrast, if all acorns germinated and grew, there would still be a developmental trajectory from acorn to oak (and savage sapling mortality). So the acorn-wormfood correlation depends on the correlation between acorn and oak, but not vice versa. One developmental path depends on the other, and hence we can regard the acorn to oak link as privileged despite its rarity. That is why it is legitimate to talk of the acorn carrying information about the tree.[2] There is not just covariation between signal and source; the genes have the biofunction of guiding phenotypic development.[3]

Our account of representation involves an important difference between our own programme and that of Dawkins. He has argued that because genes are the *beneficiaries* of adaptation, they are not *for anything*; they are not themselves adapted (Lloyd 1992). They have no teleofunctions, they just are. If that were right, a mutation would not be a mistake, only a change. And if a phenotype developed abnormally because of that change, we could not say that the gene was failing to do what it is supposed to do. Only things with functions can malfunction. So Dawkins's conception seems to rule out this option of holding that the genes are privileged in development through their role of representing the proper outcome of development.

We disagree with Dawkins in two ways. First, the roles of replicator and interactor are not exclusive. On this, we line up with Hull (1981). He points out that genes are interactors, and are adapted for those roles. Even "neutral" genes have effects at the cellular and subcellular level, effects in virtue of which they are copied.[4] Outlaw genes carry adaptations for their own replication which subvert others' prospects; for a recent example, see Werren (1991). The dual role of genes both as replicators and as bearers of adaptation is even clearer in genes which do have phenotypic effects. As a consequence of its structural and relational properties, a gene can have the function of telling the developmental programme how to build haemoglobin molecules – for that function derives from its evolutionary history (Millikan 1989). Those structural and relational properties are properties of the genes in that lineage – they are there because they have often enough resulted in those genes initiating developmental sequences that lead to normal haemoglobin, resulting often enough in the replication of that gene. Hence, there has been a malfunction if disruption of the gene's structural and relational properties leads to the formation of non-standard haemoglobin. The gene is not doing what it is supposed to. It is not doing what its ancestors have done in the past that ensured their replication.

Second, even if genes and other replicators were not themselves the bearers of adaptation, they are the products of copying mechanisms. That is, there are mechanisms which have the function of ensuring cycle to cycle similarities; of copying the replicators. So there is an error when they mistranscribe. If there are mechanisms in a bird lineage which are there because they have ensured (often enough) nest site fidelity, a copying error has taken place when a bird returns to the wrong site. The same is true of a genetic mutation: that counts as an error because a copying device has malfunctioned. Hence we can speak of misinformation in the replicator to replicator cycle.

III.3   The Varieties of Replication

We think this is the right way analysis of genetic information. But *genetic* bridges across phenotypes are not the only mechanisms of inheritance. This role of the genome is distinctive but not unique. There are a range of mechanisms through which the similarity between successive developmental systems in a lineage is maintained, and maintained as a consequence of design. Extra-genetic causal and informational transmission is not an odd footnote; it is central to the development of the phenotype. Our conception of the replicator is expansive but not promiscuous. Not every reliably re-occurring factor is a replicator. The human hand is not a replicator. The hand's biofunction is economic, not developmental. Replicators are devices with developmental biofunctions. These of course include DNA in the gametes but also a good deal else. Examples of the rest include: Kakapo track-and-bowl systems, nest site imprinting and other mechanisms of habitat stability; song learning, food preferences, and other traditional examples of cultural transmission in animals; gut micro-organism transmission in food and other micro-organism symbionts which parents are adapted to transplant to offspring; and centrioles and the other causally active non-genetic structures that accompany genetic material in the gamete. Thus, Clayton and Harvey (1993) have illustrated the heritability of nest structure. Keller and Ross (1993) have documented the cultural transmission of queen morphology in fire ants. Goodwin (1989) and Wagner (1988) detail examples of non-genetic transmission of cell structure.

Some of these mechanisms have been entirely ignored by those who focus on genetic replication. Some have been relegated to the exceptional category of cultural transmission. Still others have been taken to be just side effects of the genes. In our view they are all routes across the generations that exist, in part, in virtue of their role in ensuring parent-offspring similarities.

Indeed, we think there is a case for claiming that extending our census of the replicator makes the notion of information less mysterious. Platypus DNA bears information about future platypuses and their burrows, but only in the rather subtle sense that platypus DNA correlates by design with platypus traits. So replicators carry two sorts of information. First, they carry information about the interactions in virtue of which a new generation of replicators are constructed. Secondly, they carry information about that next generation of replicators. Non-genetic replicators may carry this second breed of information in a more direct way than genetic replicators. The construction of gene to gene links in a lineage is complex and indirect. Nongenetic replicators may bear information to rebuild similarities across generations in a more direct way: they act as templates in the construction of a new developmental system. For example, Moss argues that intercellular structures act as templates in cell division (Moss 1992, p. 345). If young platypuses tend to copy their natal burrow so that changes in a burrow are transmitted to the next burrow generation, platypus burrows may be templates for future burrows. Platypus burrows also carry information of the first kind. Obviously, they provide generalised support for platypus development. That may be all some genetic replicators do. But burrows may also be specific causes of interaction: for example, if natal burrow influences platypus choice of size, site and materials for future burrow building (on platypus biology, see Grant 1989).

Other replicators thus play the same basic role in development as genes. Nor is there any quantitative criterion that singles them out. Two obvious candidates are directness and accuracy. Hull used to defend the idea that genetic replication is particularly direct and that directness mattered (Hull 1981). Griesemer (forthcoming) has argued to the contrary, emphasising the extraordinary complexity of DNA replication. We think he is right, but like him think it is still more important to see that directness is of no evolutionary importance. It does not matter whether replicators are organisms, traits, genes, or developmental systems. The number of steps in the process through which one replicator makes another is itself of no evolutionary significance. (Smith 1994). Indeed, DNA replication is highly indirect, we conjecture, to ensure high fidelity replication.

Fidelity is of evolutionary significance. But we doubt that fidelity distinguishes genetic from other forms of replication. Other inheritance mechanisms have not been sufficiently conceived of as inheritance mechanisms for their fidelity to have been calibrated. But the problems are not just empirical. Replication could not require the reproduction of every property of the original in the copy. Only numerically identical objects share all their properties, so then we would have one replicator, not original and copy. Replication – even perfect replication – requires only the reproduction in the copy of the *relevant* parental properties. So not every change from parental song is a failure to replicate that song. Some variations are neutral. We need to understand the message before we can count variations from it. But *comparing* replication fidelity across different media is harder yet. Even when we have a sense of the message, we may not have a common currency for comparing the replication accuracy of gut micro-organisms from parent to offspring with the replication of nest site preference or DNA. We do not think information theory will help here for we cannot use an information theoretic notion of accuracy without a principled conception of the range of possibilities at that source. But what is the space of possible bird songs? Or is the space not of possible bird songs, but of possible mechanisms of species recognition or devices of territorial display?

In his 1982, Dawkins appealed to fidelity to argue that asexual organisms are not replicators (p. 97). An aphid that loses one of its legs will still give birth to six-legged offspring. If it changes, it does not pass the change on, hence is no replicator. The idea is that *a certain sort* of hi-fi is required for replication. Some copying errors are permitted; some aspects of a replicator's structure may not make it to the next link in the lineage. But any *change* in a replicator must make it to the next link. This criterion backfires against genetic replication. Many changes in the germline genes are not passed on. The point of the proofreading and repair mechanisms is to avoid the transmission of changes. So if genes are replicators, some changes in replicators need not be passed on; those censored by the proofreading and repair mechanisms. But then we can see the production of a six-legged aphid from its eventually five-legged forebear as a triumph of the *aphid's* proofreading and correction mechanisms. In any case, Dawkins's hi-fi criterion will not cull out all our candidates for non-genetic replication. A change in a bird's gut fauna may well be passed on to its descendants. The extraordinary concordance between an aphid clade and some of its symbionts (Morgan and Baumann 1994) shows such changes have been passed on, as the aphids and their fellow travellers speciated together.

Copy fidelity is relevant to evolutionary concerns. But there is no fidelity threshold that all replicators must meet. For there is a relationship between fidelity and the strength of selection. If copying is very error ridden, and selection is weak, then noise can swamp selection, and cumulative selection will be unable to build complexly adapted interactors. But stronger selective regimes can drive evolution in less perfect replication regimes (Wimsatt 1981).

So our conception of replication and replicators must enable us to identify the links in a lineages of replicators exposed by their associated interactors to cumulative selection. This conception requires perfect fidelity of neither replicator nor interactor. Indeed, the interactors associated with successive links in a lineage of replicators can vary quite widely, as adaptive plasticity, variation within a population, sexual dimorphism, and the alternation of generations in plant lineages illustrate. There is no general reason deriving from the importance of very high copying fidelity for supposing that genes are the only, or almost the only, replicators. The issue is empirical. Are other mechanisms of inheritance from interactor to interactor so noisy that cumulative selection on them is not possible? We doubt it.

Experience suggests that our extension of the role of replicators will be seen as a rejection of Weismann in favour of Lamarck. It is important to short-circuit this misconception. Weismann and his successors have shown that one apparently possible mechanism of evolutionary change is not actually possible. An organism's phenotype can change and change in a way that alters its genotype: a mouse can shift its residence to a leaky nuclear reactor. But an organism's changes cannot restructure its genotype so that its descendants manifest the changed trait. A mouse that acquires the ability to exploit a new food source cannot transmit that ability to its descendants via changes in its genome. The discovery of this constraint on inheritance is of great significance. But the discovery is a discovery of a constraint on a specific mechanism of inheritance, not a constraint on any mechanism of inheritance. Moreover, Weismann did not show that the only inheritance mechanism is genetic; few deny that social learning is an inheritance mechanism. Nor did he discover that non-genetic inheritance is subject to the same constraint. Nothing in this paper is inconsistent with Weismann's constraint on genetic inheritance. Nothing we say is inconsistent with what Weismann actually discovered rather than with what he is occasionally imagined to have discovered.

There are important distinctions amongst the elements of the developmental system. Amongst the factors that influence development, some, but not all, are part of a copying and interaction cycle. Garbage cans are part of the developmental matrix of many suburban-adapted Australian possums, but possum behaviour does not result in a flow of new cans. Only some elements of the developmental matrix are adapted for their role in development. The explanation for their existence and nature is that earlier copies played a similar role in the development of similar phenotypic systems. Platypus burrows exist in their contemporary forms because earlier copies played a similar role in the development of burrowing platypuses. That is not true of the relationship between the paradise parrot and the termite mounds they nested in. Termite mounds continue; sadly, the paradise parrot does not.

IV   REPLICATORS AND INTERACTORS

Most of the original motivations for the replicator/interactor conception are really arguments against the Received View rather than arguments for a particular heterodoxy. But not all: one interesting and important line of thought supports the replicator/interactor conception of evolution.

IV.1   Organismism

Hull and Dawkins use the Gene's Eye conception of evolution to release the grip of the organism on our biological imagination.

Hull emphasises botanical and other examples that undercut the idea that the organism is well-defined and readily identifiable (1988a, Chapter 11). In his hands, this line of thought emphasises the atypicality of the "paradigm organism": the multi-cellular animal. So conceived, organisms are rare, special, aberrant. Most of life does not come in packages like that. Protozoa, colonial organisms, many plants and their clones do not fit. Evolutionary theory should not be conceptualised by appeal to examples that are atypical of life's evolution.

Hull reinforces these intuitive considerations by a more formal argument. The natural kinds of biology are those that play a distinctive role in biological theory. There is nothing, Hull argues, that all and only genes, organisms, groups or species do. Nothing of evolutionary interest is true of, for example, all and only organisms. The natural kinds are replicators, interactors and lineages. Genes are paradigm replicators, but not quite the only ones: in asexual or genetically homogeneous populations (e.g., cheetahs) much larger units – chromosomes, genomes or even the organism itself may qualify. Organisms are the paradigm interactors but not the only ones.

Dawkins's arguments for the extended phenotype (1982 and 1989) are another version of this idea. Selection does not see the naked replicator. Genes are selected in virtue of their phenotypic effects. But the effects in question need not be effects expressed in the body in which the replicated gene is situated. Genes have extended phenotypes. Some of their jointly constructed adaptations are aspects of the organisms in which they ride, and through which they replicate. But some are not. Dawkins emphasises adaptation-at-a-distance: adaptations that aid the replication of genes, but which are not adaptations of the body in which the genes live. Adaptation-at-a-distance enables us to see the evolutionary identity of nest building of the caddis fly with shell secretion of molluscs. Though the caddis house is not a trait of the organism, it is just as much an adaptation, and an aspect of the caddis genes' active replication. The same perspective shift enables us to see the altered behaviour of a parasite's host as an adaptation of the parasite's genes.

This line of thought is permissive rather than compelling: the Received View remains viable. It is possible to insist that the adaptation in question is the secretion altering the behaviour of the host. But there does seem to be reason for singling out one link in this chain. The effects on the hosts' behaviour is the most salient link.

71

The adaptive effect of the parasite's genes is the effect on the host's behaviour.

We like these arguments. But they are neutral amongst those positions which reject the Received View. Consider, for example, a robin feeding a robin chick and it feeding a cuckoo. On the Received View, one comes out as adaptive behaviour; the other, maladaptive. But the alternatives capture a commonality. Developmental Systems Theorists see both behaviours as stable results of typical packages of developmental interactants which tend to recreate the conditions for their own reproduction. In both the robin-robin and the robin-cuckoo systems, the boundaries of the organism are of no special significance. With this the defenders of the Gene's Eye agree: both behaviours express extended phenotypic effects. The feeding genes are in the chicks. The robin is being manipulated by outside genes in both cases, but we overlook robin chick manipulation because we are often blind to conflicts of interest between parent and offspring. The Extended Replicator, in this instance, takes over the Gene's Eye explanation. For us though, it is an open question whether the genes are the only replicators in these interactions with extended phenotypic effects. In the manipulation of the robin, other replicators may play a role. The physical surrounds – the nest structure – may be an essential feature of the chick's manipulations. So the nest may be collaborating with the robin chick (and be cheated by the cuckoo chick) to ensure a new generation of nests (see V.1). There may be egg-line but not DNA factors essential to the survival of the cuckoo's egg, or the development of the chick's manipulative skills. In sum, the heterodox will describe these cases in somewhat different languages, but they all see something that the Received View misses.

IV.2 Comparing Heterodox
Perspectives

Dawkins (1982) points to phenomena that can be explained from the perspective of the Received View, but which can be much more easily grasped from that of the Gene's Eye. For example, it is much easier to see the problem posed by sexual reproduction from the gene's perspective. How can it benefit a gene to collaborate in a system of replication in which its chances of being replicated are only 50 percent? The *Extended Phenotype* is full of examples of the value of these perspective shifts. One particularly striking example is that of the organism itself. From the perspective of the germline cells, the construction of the body is an enormous investment of resources that might instead be directed directly to replication. Why is it worth it? Why is this investment not inevitably subverted by cell-line rebellion (Buss 1987)? A perspective shift reveals these questions. We see no help here from the Developmental System perspective. On that view, neither the existence of organisms nor of sexual reproduction seem particularly problematic. No doubt it is possible to formulate these problems in that language, but they are not "in your face."

So one important argument for the Gene's Eye is not undermined by the more radical perspective. Since we take the genes to be replicators, we inherit this advantage. But why prefer our picture to that of Dawkins? There is a formal argument: any decent definition of replication applies to lots of non-genes, for there are many inheritance mechanisms, not just one or two. There is a historical argument: genes evolved from earlier extinct replicators (Cairns-Smith 1982). We think there may also be a heuristic argument: seeing certain processes as replication opens up new questions and poses interesting problems. We cannot claim proof here: it remains to be seen whether that perspective generates fruitful new work. But we think it may, while blocking none of the heuristic advantages of Dawkins's perspective shift.

Consider, for example, sexual reproduction. Sexuality surely says something about differences amongst developmental agents. Genetic recombination within a restricted group produces controlled phenotypic diversity. If other developmental factors are capable of playing the same role, we would expect to see reproductive systems that exploit other factors to generate but control diversity. So we expect to find asexual lineages which generate controlled phenotypic diversity by varying the developmental matrix through migration, or through varying the time or season of reproduction. And we expect to find sexual lineages where controlled diversity is generated not just by recombination but by other adaptations. Our hunch is that there are other elements exploited to generate diversity and that these will be adapted to play a similar role to germline genes. If we turn out to be right, then our conception would be shown to have the same heuristic advantages over the Gene's Eye that it has over the Received View. Moreover, it would decisively rebut Sober's claim (e.g., Sober 1990) that if gene selectionist claims are about replication they are trivial, merely telling us what we already know.

IV.3   What Is Copying?

We need an account of replication which does not prejudge the count of replicators, nor understate the complexities of replication. We do not see complexity as problematic, for an e-mail copy of this chapter is very indirectly produced, and depends essentially on many elements additional to the Word document. Yet it is a copy for all that. So without attempting an explicit definition, we propose that the following elements are part of the biologically interesting notion of replication.

If B is a copy of A:

(i)   A plays a causal role in the production of B.

(ii) B carries information about A in virtue of being relevantly similar to A. This similarity is often functional: B. has the same, or similar, functional capacities to A. Indeed, we should probably think of "copy" as a three-term relation: B is a copy of A with respect to C, where C is often some function of A.

(iii) B respects the xerox condition: B is a potential input to a process of the same type that produced it.

(iv) Copying is a teleological notion. For B to be a copy of A it must be the output of a process whose biofunction is to conserve function. On this view, the mere similarity of B to A does not suffice for B to be a copy of A. A fossil of a leaf is not a copy of a leaf. B must be meant to be similar to A; that similarity is why those mechanisms exist. Copying is a process with the function of producing from one token another which is relevantly similar.

We can thus see why genes can be copies without downplaying the complexity of this process. Nor need we suppose that genes, nor any other replicator, are "self-replicating."

V   AGAINST LIFEISM?

There are many mechanisms of inheritance. In this final section, we wish to explore the possibility of a dramatic extension of inheritance notions by examining recent rhetoric in evolutionary biology. One strand of the anti-adaptationist literature, and one strand of vicariance biogeography, have talked of life and the environment evolving together. We take seriously the idea that elements of the environment evolve.

V.1   The Selfish Burrow

Nesting burrows are replicators. The causal relations between burrows and burrowers is like that between genes and their interactors. No gene makes an organism. But variance explains variance: a variable oyster-catcher may be black rather than pied because it has one gene complex rather than another, even though no gene complex makes a colour pattern. Similarly, a variation in a burrow can cause a variation in a burrower: a particular penguin chick may be healthy and safe because its burrow has one site rather than another, even though no burrow features make penguin flesh.

Second, burrows are part of a copying and interaction cycle. They exist in the forms they do because of their role in this cycle. We think this is important. Griffiths has urged against our extension of the concept of replicator that we underplay the significance of relationships and their reproduction. So he points out:

> the evolution of the hermit crab-shell relationship is interesting. Something is being replicated, and it isn't the shell. Is it the shell-using behaviour? Maybe, but that would hardly explain the evolutionary dynamics of the population documented by Gould which uses fossil shells and is now going extinct. (personal communication)

We read a quite different moral out of this example. The crab-fossil shell relationship is not a replicator, *precisely because* the hermit crab is unable to influence the availability of a critical resource in the next generation. There are no mechanisms in this developmental matrix which have the biofunction of shelter-making. That is part of the rather sad evolutionary dynamics here. So while this relationship is of evolutionary interest, it is not copied, though shell hunting and occupying behaviour may be. The hermit crab snail shell relationship is thus quite unlike that between penguins and their burrows.

Burrows bear information about burrowers and the next burrow generation. For burrows interact with their guests in ways that result in mutual changes. They and their guests co-evolve. Chance changes in burrow copies can proliferate. A chance favourable burrow copying at a new and superior site – for example, one less liable to flood – may result in a bushy lineage of similarly changed burrows. So they are replicators in the sense closest to Dawkins's heart: a change in a burrow is sometimes copied through to the next generation.

These ideas emerge most clearly from a consideration of Bateson's challenge to *The Selfish Gene*. Bateson (1978) pointed out that one consequence of Dawkins's conception is that nests are replicators, and that a bird is but a nest's way of making another nest. Dawkins rejected this idea on the grounds that variation is not transmitted. Whatever the merits of the Selfish Nest as an evolutionary hypothesis, it cannot be rejected on those grounds. First, because Dawkins appeals here to the same criterion used to exclude asexual organisms as replicators; a criterion unsatisfactory on other grounds. Second, it is not in general true. Environmentally altered patterns in cilia are inherited through fission (Majerus and Hurst 1993). Variation in both nesting materials and nest siting can be transmitted (Dickison 1992; Gray 1992). Perhaps even variations in builders can be: some New Zealand petrel nests are inhabited by both petrels who build them and tuatara who live in and maintain them all year around: this change in occupancy pattern has become typical of that local deme of burrows, and is now part of local burrow-genesis. Naturally, as in all new fields, delicate empirical questions remain. Tuatara are known to eat nestling petrels from time to time. So it's a question for future research whether this impact on the burrow-building population so adversely affects burrow fecundity as to outweigh their positive effect on longevity. If so, we would have to regard tuatara as burrow-parasites.

If nests are replicators, they are clearly active replicators. Their properties of insulation, durability, protection and cost of construction quite obviously influence the probability of their being reproduced. They form lineages. A nest plays a role in the construction and protection of builders who disperse to produce a new and relevantly similar nest. There is a flow of information linking nest generations through the builders. Nests and burrows are adapted for the growth of burrow builders and nest-makers. Those interactors carry the information through which the nest is replicated.

Bateson was right: sometimes, perhaps always, nests meet Dawkins's definition of a replicator, and they meet ours too. Nests are produced by mechanisms whose biofunction is to reproduce this critical developmental resource. But there is no reason to suppose this is a reductio of these accounts of replication; Dawkins here betrayed a rare moment of timidity.

V.2 Selfish Burrows or Selfish
Burrowing Genes?

Of course, a Gene's Eye conservative may deny that the burrow is the replicator here. They might claim that the real replicator is the penguin gene or meme for burrowing on (say) predator-free Kapiti Island rather than the predator-infested Wellington foreshores.

Redescription is possible but not appropriate. First, it is certainly possible for the burrow lineage to grow bushy without a burrowing meme that grows bushy. The successful colony can grow as the result of greater than average burrow success, even if burrow residents are no more likely to reburrow there next season. The increase in burrowers together with the differential survival of Kapiti burrows might explain a bushier burrow lineage there: its growth does not imply a change in the genetics or psychology of its denizens. Moreover, the considerations which favour Dawkins's conception of an extended phenotype reapply here. With the usual caveats about costs, changes in the developmental mechanisms that make no difference to the caddis house itself are selectively neutral; only variations that vary house design matter. Hence, it is reasonable to focus on caddis houses rather than caddis larvae house building. Similarly, changes in the way burrows are built that do not result in variation amongst burrows are not relevant to burrow evolution. It's the burrow, not the particular form of the burrowing meme that matters to burrow evolution.

Second, we can accept that burrow genes and burrowing memes are replicators without denying that burrows are replicators. There probably will be a site tradition handed on amongst burrow denizens, so as the burrow lineage grows bushy this one will grow bushy with it. The Extended Replicator will often allow a two-way view. From the perspective of the burrow, the burrower is an interactor whose differential reproduction differentially replicates the burrow. Equally, from the perspective of the burrowing genes and memes the burrow is an interactor – part of the interaction process whose differential reproduction promotes a longer and bushier burrowing gene. The same will be true of more conventional mutualisms; for example those between ants and plants. From the perspective of each the other is an interactor: adapted to the environment in such a way that its success vis-à-vis its competitors results in the differential replication of its colleague.

There is nothing mysterious about this perspective in which the one entity acts both as replicator and interactor; both the beneficiary of adaptation, and the bearer of adaptation (Lloyd 1992). The Gene's Eye view is itself committed to a restricted form of this pluralism. Even if we were to restrict our focus entirely to genetic replication, not all of a gene's adaptations need be adaptations for that gene. Sometimes they are: because they ensure its replication, various molecular and cellular relations of a gene will count as adaptations of it for it. But from the perspective of other genes, these can be interactors carrying their adaptations for other genes' benefit. For example, consider a repressor gene. For other genes, it's an interactor carrying adaptations for their benefit. The characteristics it has to ensure high-fidelity replication benefit them too. They may well themselves have characteristics which make its replication and repression more effective (Moss 1992). So some of their characters count as adaptations they bear partially for it. It benefits from their characters; they benefit from it. Now consider an ex-outlaw gene now permanently turned off, and hence part of junk DNA. It still gets replicated, so some of its structures and relations ensure that, and hence are adaptations for it. But some of those relations ensure it is inactive. Though these are adaptations of the outlaw gene, they are not adaptations for it but for other genes. They are extended phenotypic effects of, and adaptations for, repressor genes.

Third, there is no reason to privilege the causal dependence of the selfish burrow on selfish genes and memes. Of course, it is true that there is only a causal chain from burrow to burrow because there are causal chains from burrow genes to burrow genes. The burrows can replicate only if burrow genes replicate. But this dependence is symmetrical: holding causal background constant, burrow genes replicate only if burrows replicate. If the chain of burrows were to fail, the chain of burrow genes fails too. Nor can this symmetry be broken by tracking back into history. Each replicator has been part of the evolutionary history of the other, and we are certainly in no position to claim that any current burrow gene or meme lineage is deeper rooted in history than the burrow lineages.

For those who think this view must be a joke, at best, we offer the following two cautions. First, evolution produced the paradigms of the living, so evolution cannot be restricted to lineages of entities that we would now count as living. Evolutionary change in lineages of non-living or quasi-living entities must be possible. Second, it does not follow that these evolutionary processes have the same power as those operating in the living world. It is sometimes said that one gets evolution under natural selection whenever there is heritability, variation and differential fitness amongst the variants. Perhaps so, but if selection is to explain major adaptation it must be cumulative. Innovation is the result of a long sequence of selective episodes rather than one. Cumulative selection requires much tighter conditions. So even if nests and burrows evolve, it may be that their evolutionary dynamics and possibilities are more like those of the long-disappeared lineages of proto-prokaryotic cells rather than those that have led to complex adaptation.

VI CONCLUSION

In sum, the radicals' insistence of the seamless nature of developmental interaction is important and they are right to deny to the gene any exclusive role in development or evolution. But we think both the replicator-interactor distinction, and some of the reasons for making it, survive their critique. However, rebutting the Developmental Systems Theorist's critique extends our conception of the cast of replicators, and hence should shift our perspective from that of the Gene's Eye to that of a still more raucous and motley crowd of squabbling replicators.[5]

NOTES

1. See paradigmatically: Oyama 1985; Gray 1992; Griffiths and Gray 1994; Moss 1992. Oyama prefers to speak of "Evolutionary Developmental Systems." In her view, we are interested in a species of a larger genus of views and interests. This group draws on other figures: see especially Bateson (1976; 1983; 1991). Johnston 1987 and 1988. Griesemer forthcoming and Smith 1992 and 1993a defend some of their distinctive negative theses but not their positive ones.
2. Refugees from the philosophy of mind will recognise here an adaptation of Fodor on mental content; see for example his 1990, Chapter 4. Ruth Millikan (1991) pointed out that these asymmetries must be given a teleological twist.
3. It's a consequence of our view that the informational role of the gene is to carry information about the parental phenotype to the developing phenotype. It would follow that the source is the interactor/replicator system of generation 0, the signal is the gamete, and the receiver is the interactor/replicator system of generation 1.
4. We disagree on whether neutral genes are interactors. For a neutral gene to be an interactor, it must have effects which makes its replication differential with respect to actual or past rivals. Sterelny thinks that the relational

Replication and Interaction

properties of neutral genes can give them replication advantages; for example, they can hitch-hike. Smith and Dickison think that this trivialises the concept of an interactor. On both views, a mutation in a neutral gene counts as a mistake in the second sense: a copying mechanism has not operated as designed. In Sterelny's view, a neutral gene can fail to do as it is designed to do in the first sense as well: if some disruption of its structural and relational properties block its replication.

5. This was co-authored with Kelly Smith and Mike Dickison. Thanks to Sandy Bartle, David Braddon-Mitchell, Richard Dawkins, James Griesemer, Peter Godfrey-Smith, David Hull, James MacLaurin, Susan Oyama, Steve Trewick and a referee for this journal for their comments on earlier drafts of this chapter. Particular thanks to Paul Griffiths and Russell Gray for many hours of talk and correspondence that were part of its developmental matrix.

80

4

The Return of the Group

1   A SLICE OF HISTORY

Once upon a time in evolutionary theory, everything happened for the best. Predators killed only the old or the sick. Pecking orders and other dominance hierarchies minimized wasteful conflict within the group. Male displays ensured that only the best and the fittest had mates. In the culmination of this tradition, Wynne-Edwards (1962, 1986) argued that many species have mechanisms that ensure groups do not over-exploit their resource base. The "central function" of territoriality in birds and other higher animals is "of limiting the numbers of occupants per unit area of habitat" (1986, p. 6). Species with dominance hierarchies, species with lekking breeding systems, and species with communal breeding regulate their populations. These social mechanisms have population regulation as their "underlying primary function" (1986, p. 9). Wynne-Edwards argued that these mechanisms evolve through group selection. Populations without such mechanisms are apt to go extinct by eroding their own resource base.

Group selectionist ideas were motivated by the perception of altruistic behavior in the natural world. It is hard to see how altruism could evolve by individual selection, for in mixed populations the selfish do better by free-riding on the others. But it is not hard to see how groups of altruists would out-compete selfish groups. Sadly for these visions of benevolence, group selection fell on very hard times. Lack (1966) argued that reproductive restraint *can* benefit the individual. More generally, Maynard Smith (1964) and Williams (1966) argued that once made explicit, the assumptions on which group selection relies seem very implausible. In sum, the reactions to group selection took three forms: the first was an alternative conception of evolution and the units of selection, the second was an alternative explanation of altruism and other explanatory targets of group selection, and the third was a direct critique of group selectionist explanations.

Williams, and following him Dawkins, argued that the real agents in the selection game are genes. In reproduction, some of the germline genes of an organism are copied and passed on to their offspring. Genes are replicators. They root lineages of copies of themselves. Some lineages are long and bushy, others are short or thin. For the most part, these differences are no accident: genes exert phenotypic power on the world in ways that affect their copying propensity. The adaptive features of organisms ultimately benefit the genes responsible for them, for the consequence of adaptive improvement is the proliferation of genes that rebuild those adaptations in descendant generations. The fundamental cause of selection is the differential capacities of replicators; its fundamental consequence is the differential growth of replicator lineages. On this conception, organisms are not copied. Reproduction is not the replication of organisms. Organisms are not the beneficiaries of adaptations for they cannot make more of themselves. Rather, organisms are vehicles or interactors.[1] Vehicles interact with their environment as a cohesive whole. They are structured units, units with effects on one another, effects with downstream consequences on replication propensities. Their differential success vis-à-vis their ecological competitors causes the differential success of replicator lineages vis-à-vis their genealogical competitors.

The second blow to group selection was the articulation of explanations of the most striking examples of altruistic animal behaviour that seemed to leave no room for group selection. The most striking examples of co-operation – the complex divisions of labor amongst the eusocial insects – have come to be seen as the result not of selection at the level of hives or communities, but rather as the result of kin selection. Hamilton showed that in calculating an organism's fitness we should consider not just that organism's offspring, but also the contribution that organism makes to the direct fitness of its relatives, discounted by their genetic distance. Indirect fitness benefits are central to the evolution of the spectacularly altruistic behaviours of social insects.

The critics of group selection also point to a deep problem for Wynne-Edward's picture of restrained groups out-competing profligate ones. Altruistic groups seem very vulnerable to subversion from within. Imagine rabbit populations differing in levels of reproductive restraint. Restrained groups, let us suppose, delay first breeding. Restrained groups do much better in harsh winters, for most rabbits in profligate groups starve. Even so, if there is migration between groups, or if more fecund rabbits arise easily in the population, they will gain the benefits of living in a restrained group without paying the costs. Hence, the fecund will white-ant restraint. Moreover, the faster generation time of individual selection and the greater variety of individuals, suggests that in a race between the two selection processes, individual level selection should win. No-one claimed that group selection was inherently impossible. Rather the claim was that very special conditions are needed before it works.

So group selectionist ideas fell on very hard times in mainstream evolutionary theory. But they are now making a comeback, due mostly to the work of D. S. Wilson.[2] It is that comeback I consider in this chapter. The central ideas involve: (a) the idea that group selection is a hypothesis about vehicles, not replicators; (b) reinterpreting kin selection and reciprocal altruism as instances of group selection; (c) unveiling the excessively strong assumptions implicit in those analyses that suggest group selection must be weak or rare; (d) some suggestive examples of group selection in operation. Let me unpack these cryptic remarks.

2 THE RETURN OF THE GROUP: TRAIT
GROUP SELECTION

The best way to interpret group selectionists is to see them as claiming that groups function in evolution in the same way organisms do. That is, group selectionist hypotheses are about vehicles, not replicators. So a group selectionist can agree that the only replicators are genes; they think though that some vehicles are groups. So group selection is independent of the gene's-eye conception of evolution. The idea that group selection is a hypothesis about vehicles opens the door for a revival of group selection, but is by itself insufficient. Even once we have made the replicator/vehicle distinction, group selection can seem unlikely if we have an overly fancy conception of vehicles. Wilson and his allies think our standard example – the vertebrate organism – can give us a misleading picture of vehicles. For such organisms are extraordinarily complex, co-adapted, and enduring. As Dawkins has emphasized, they have a very distinctive developmental sequence. They have

83

determinate boundaries and they act on their environment as cohesive wholes. If every vehicle must have these characteristics, then not many collectives will count.

Wilson has argued that we should not restrict our idea of a vehicle to collectives which are co-adapted and complex. His argument depends on drawing a parallel with the evolution of individual organisms. Organisms are collectives or assemblages of cells. In the paradigm cases, they are assemblages in which competition between cells has been suppressed. But Buss (1987) showed that not all cell-assemblages have a full suite of adaptations that suppress that competition. Those adaptations arise through a history of cell-assemblage competition. Organisms must have been vehicles before they were fancy vehicles. Similarly, obvious group level adaptations are products of a long history of group selection, not preconditions of its beginning.[3]

Wilson and friends therefore propose to identify a vehicle through the idea of common fate. An organism is a vehicle rather than a population of cells because those cells share a common fate. Similarly, a group of beavers in a dam is a vehicle if their fitness is linked together on a common causal trajectory. Beaver traits that affect that trajectory for better or worse can be visible to selection through the fate of that beaver collective. Obviously, common fate comes in degrees. No-one suggests that the fate of the beavers is as interconnected as the fate of the cells of one beaver. More importantly yet, common fate is defined on a trait-by-trait basis. Beavers, unlike cells in a single beaver, vary in fitness. Still, if the beavers co-operate in the construction and maintenance of their dam, then that characteristic will have a common effect on the beavers in the dam. Those beavers will be affected and in the same way by this feature of its members, and hence with respect to this trait – the dam-building trait – the beavers share a common fate.

Thus Wilson introduced and defended the idea of "trait groups": populations which have a common fate with respect to some trait. Trait groups are groups of organisms each of which feels the influence of the others with respect to some trait. So if the trait is that of making warning signs, it is the group of beavers that are in earshot of each others' tail slaps. If the trait is dam building, it is the group of beavers that live and shelter in the dam. Different traits will pick out different groups. In the most obvious examples, these groups are homogenous with respect to the trait in question. So our trait groups would be beaver groups in which all the beavers can hear each others' signals;

or the group of beavers all of whom live in the dam they maintain jointly:

> ... every trait has a "sphere of influence" within which the homogeneity assumption is roughly satisfied. It is the area within which every individual feels the effects of every other individual. I have termed this population the trait group ... to emphasise its dependence on the particular traits being manifested." (Wilson 1980, p. 22)

However, trait groups need not be homogeneous with respect to the trait whose evolution is of interest to us. Kin groups of mixed altruists and non-altruists are a single group, as are mixed groups of co-operators and defectors in social interactions. So a trait-group playing a role in the evolution of T is not necessarily a group of organisms all of whom have T. It is a group of organisms all of whom are roughly equally affected by the same group of T-bearers. All the beavers that live in a dam are a trait group, even if a few free-ride. For all these beavers, and only these beavers, are in "the sphere of influence" of these dam builders even if they are not all themselves dam builders.[4]

A group of beavers in a dam is a spatially distinct group. But trait groups need not be spatially distinct in this way. A beaver that warns only blacktailed beavers forms a trait group: the group of beavers in the "sphere of influence" of the bearers of that trait. It does so even if these beavers are not spatially segregated from the rest. The fact that trait groups need not be spatially segregated enables us to see kin groups as trait groups. If trait groups are vehicles, kin selection is not a superior alternative to group selection, but a version of it. For altruism towards kin generates trait groups, the organisms all of whom are potential beneficiaries of that aid. This division of the population into groups is essential to the evolution of altruism. Within the kin group, the defector does better than his altruistic sibs. For the defector enjoys the benefits of aid without bearing the cost of giving it. A female lioness that did not allow her sisters' cubs to suckle would improve the prospects of her own reaching maturity. But a pride of altruist lions will "fledge" more cubs than a mixed group, which will do better than a wholly defecting pride. So if the boost from altruism is big enough, and the cost is small enough, the average fitness of the altruist can be greater than the average fitness of the defector. Hence, when the kin groups dissolve back into the general population before the next round of breeding, the proportion of altruists can rise despite free-riding in mixed groups.

Kinship is important, for it plays the central role in generating variation between groups. Since kin resemble one another, kin groups with one altruist are likely to have more than one; similarly, with selfishness. So the traits in question are not distributed randomly across groups. Random distribution, of course, depresses variation.

The same reasoning suggests that the evolution of reciprocation, too, frequently involves vehicles composed of individual organisms. Wilson and Sober (1994) unveil a most striking thought experiment with this as its moral. They depict a cricket population that feeds on lilies scattered across a pond. The problem for the crickets is to get from one lily to the next. Wilson and Sober imagine the evolution of co-operative navigation across the pond, as pairs of crickets evolve the capacity to row between lily pads on dead leaves. In their view, the required co-ordination evolves by selection on pairs. Crickets better able to co-ordinate with their partners are fitter than their clumsier colleagues. But this adaptive advantage is visible to selection only through the increase in efficiency with which *a pair* reaches a lily pad. So the pair is a vehicle. With respect to each trip, the pair share a common fate, and hence co-ordination evolves by group selection. The pair is a vehicle, even if these are the only co-operative interactions between crickets; even if a cricket rarely has the same partner twice; even if the great bulk of the cricket life cycle is between trips.

Evolution continues, as evolution will. A selfish mutant arises which casts adrift its partner at the end of the trip. It does well when its partner is naive, but poorly when paired with another selfish morph, for each has a tendency to drown the other. Despite group selection in favour of co-operation (for co-operative pairs do better than selfish pairs) within-pair selection favours the mutant and causes the selfish behaviour to spread. More evolution: eventually a suppresser morph arises that prevents the selfish morph's behaviour by clasping it when the two arrive at their destination. The clasping morph spreads through the whole population by group selection alone, for whatever the nature of its partner, the two crickets benefit equally from every trip. There are no within-pair fitness differences. Throughout this whole evolutionary dynamic, Wilson and Sober think the pair is a vehicle. It is the beneficiary of the joint behaviour whose benefit is distributed over its members.

This picture of water cricket evolution suggests that the "subversion problem" is much overstated. That problem depends on the idea that groups are fairly large, they persist for a number of generations, and they are unlikely to have effective defences against subversion.[5] If water cricket evolution is evolution by group selection, these assumptions are unwarranted.

Even those very sceptical about group selection have always agreed that there might be a few examples of group selection in action. So, for example, Wade and his students were able to show that group size in flour beetles responds to selection (Wade 1978). More important, though also more controversial, was the evidence for the role of group selection on sex ratios. Selection at the level of individual organisms tends to favor a 50/50 ratio. For if the sex ratio is disturbed from a 50/50 balance, a parent maximizes its grand-offspring, all else being equal, by producing the rare sex. However, it is easy to conceive of circumstances in which selection on groups would favour an unbalanced sex ratio. Imagine insects that feed on rich but widely scattered resources. The resource is typically found by one or a few mated females, whose progeny consume the resource. They mate and the females disperse. Almost all die, but a few find new bonanzas. The more that disperse, the more likely it is that there will be some successful searchers. A female-biased sex ratio allows more colonizations. Under circumstances like these a female-biased sex ratio can evolve, and has evolved, apparently many times. Both Williams and Hamilton accept that female-biased sex ratios in nature are good evidence for group selection on sex ratios.

However, if Wilson and his allies are right, all this is small beer. In the last few decades, evolutionary analyses of social behaviour have depended on two central ideas: kin selection is one, evolutionary game theory – most obviously, reciprocal altruism – is the other. Trait group theorists see most of this work as implicitly depending on group selection. So not only are the conditions on group selection less restrictive than had been proposed, but more important, mechanisms which all agree are of great importance in social evolution rely on selection between trait groups. Famously, vampire bats share food. These bats die unless they feed every couple of days, and hunting failure is quite common. So food sharing is essential to vampire life. Bats that give are bats that receive. Here, too, we have a trait group, and the vehicle is the bat population that roosts together. These groups have a common fate. Punishment of defection – of bats that fail to share – suppresses the free-rider problem in ways that parallel the water cricket's clasping adaptation.

If reciprocation is important, kin selection is even more important. The complex social adaptations of honeybees and other highly social insects are well known, and no doubt kin selection has played an important role in their evolution. Kin selection is important not just in these well-known cases. One of their most striking examples is of a parasitic worm (Sober and Wilson 1998). The trematode parasite *Dicrocoelium dendriticum* has a life cycle which takes it through three separate hosts. Adults live in livestock; they lay eggs in the livestock's dung. These eggs are eaten by snails, in whom they hatch and in whom they reproduce asexually for two generations before forming a mucus-covered larval mass which the snail excretes. This mass of several hundred parasites is eaten by ants, the worms' next host. At this stage one of the worm larvae invades the ant's nervous system and changes the ant's behaviour so that it spends much of its time on grass tips, thus much increasing the chances that the ant will be eaten along with the grass. Should this happen the brainworm dies, but promotes the completion of the life cycle by the other larvae.

In the picture of evolution informed by trait group selection, the division of organisms into trait-defined groups plays a significant role in the evolution of a wide range of behaviours in many different clades. These theorists do not expect trait group selection to be the only force acting in the evolution of some trait. Often the population of groups will include "mixed" groups, and the outcome will depend on the combination of selection between groups and selection within groups. The view they reject, though, is the idea that we should standardly expect within-group selection to swamp between-group selection.

3 REACTIONS TO TRAIT GROUP SELECTION

Trait group selection raises three distinct issues. First, granted that trait groups are important aspects in many evolutionary processes, are trait groups vehicles? An alternative is to think trait groups are a critical part of the environment in which individuals evolve. Second, this possibility raises an overarching question. Is there a single best way of describing these evolutionary episodes (Wilson's way) or, rather, are there a number of equally adequate descriptions of evolution involving trait groups? Dugatkin and Reeve (1994) defend the pluralist idea that there are a number of equally adequate depictions of the trait group phenomena. And third, the pluralist option returns us to the relationship between trait group selection on the one hand and traditional group selection on the other. Wilson and Sober clearly see their work as a partial vindication of the Wynne-Edwards tradition. But that is not obviously so. Wynne-Edwards really did see groups as organism-like. He saw some groups as complex co-adapted wholes. Yet, as we have seen, trait groups need not be much like organisms. The relationship between trait group selection and its ancestors is of more than scholarly interest, for I think that two conceptions of group selection co-exist in the literature, and that trait group evolution characterizes a weaker notion of group selection than "superorganism" evolution.

3.1   Groups: Vehicle or Environment?

The division of a population into groups surely is important in evolution. But it does not follow that these groups function in ways that parallel the role of organisms. Dugatkin and Reeve argue that there is an alternative account of the evolutionary processes on which Wilson and his allies focus. On this alternative view – "broad individualism" – groups are important aspects of the environment in which selection occurs.

Consider first an example for which the group selectionist perspective is very persuasive. Social insect groups really do seem to be vehicles. They are cohesive, co-adapted, and share a common fate. Ant nests and bee hives seem just as "visible" to selection as individual ants and bees. It is surely just as likely that an ant gene promotes its replication by its effect on an ant nest as on the individual ant in which it resides. Many species of ant have elaborate chemical and behavioural warning and defence mechanisms. Often individual worker's defence chemicals serve a second function of recruiting aid and alerting the nest to danger. Perhaps the genes for these mechanisms replicate via their effects on the nest. Nests with better warning systems last longer and found more colonies, all else being equal, than less efficient nests, and hence the warning genes are preferentially replicated in the ant gene pool.

Even for cases like these, there is an alternative story. Ants carrying the warning gene are fitter in environments in which other ants carry the same gene. In those environments, their signals will attract the appropriate response. Their own responses, too, will be appropriate, for they will not be alone in responding. The close genetic relationship within the nest makes it likely that an ant with the warning gene will be in similar company. So instead of seeing the warning gene as evolving via selection between ant nests we should see it as selection between individual organisms. Ants with the warning gene are fitter, on average, than ants without the warning genes. As always, this fitness advantage depends on the environment in which this evolutionary change is taking place. In this case, a key feature of the environment is the population structure of the ant population itself: the warning gene is fitter only because the population is subdivided into nests each of which consists of close genetic relatives. Nests turn out to be a key feature of the *selective environment*.[6]

A similar line of thought challenges Wilson's and Sober's reinterpretation of kin selection. In environments in which altruists are mostly segregated into groups with others like them, on average altruists will be fitter than defectors, even though the very fittest organisms are the necessarily exceptional free-riders. Again, groups of organisms play a key role as part of the selective environment, not as vehicles of selection. On this way of telling the story, genes for female-biased sex ratios have adaptive consequences for their bearers in certain distinctive social and ecological environments.[7] The moral extends to reciprocation. Instead of thinking that water cricket evolution is driven by group selection, think instead that it is driven by frequency-dependent selection on individual crickets, selection driven by the relative frequency of different types of cricket. Thus the clasping gene is adaptive only when defectors are common. Apparent selection for co-operative groups is selection for co-operative individuals provided there are enough other co-operators in their environment. So group selection is converted into frequency-dependent selection on individual organisms.[8] Applying this idea, we could reanalyse the evolution of (say) large flour beetle groups as the evolution of a gene for (say) synchronising breeding, a gene advantageous only in a particular population-structured context.[9] Once more, groups become part of the selective environment, not the vehicle through which the synchronising gene ensures its replication.

3.2 One True Story?

In his 1987, Maynard Smith suggests that trait group selection is quite often equivalent to broad individualism.[10] He does not think this equivalence is complete, for he thinks that kin selection need not involve a population being segmented into groups. Dugatkin and Reeve (1994) go the whole hog in defending pluralism. They do not claim that broad individualism is the right picture and trait group selection wrong. Rather, they argue that each is equally good; we have two equivalent descriptions of (say) the evolution of honey bee altruism. On only one the colony counts as a vehicle. On this view, while the caste of replicators is an objective feature of the history of life, the caste of vehicles is not. There is no fact of the matter as to whether a vampire pair that shares blood is really a vehicle of selection or not. There can, however, be heuristic differences between the two approaches. So the vehicles we should recognise depends on our explanatory and predictive interests.

A similar moral emerges from one way of reading Dawkins' reaction to trait group selection. Dawkins (1982) defends the idea of the "extended phenotype." Genes have effects that promote their own replication but which are not expressed as traits of the organism in which they ride. So extended phenotype genes are active replicators, but not replicators that compete with their rivals by constructing vehicles. Genes that code for kin-selected traits and reciprocation are genes with extended phenotypic effects. So Dawkins (1994) rejects the idea that kin groups and reciprocation groups are vehicles. But instead of accepting broad individualism, he argues that in the evolution of water cricket navigation there is no vehicle at all.

Amongst Dawkins' most striking examples are parasite gene effects expressed as host behaviours, and this raises the possibility that trematode altruism is a consequence of the manipulation by genes in other larvae. Perhaps there is a trematode arms race, with all the larvae exuding both "turn into brainworm" signals aimed at their fellows, and smokescreens to protect themselves. The brainworm is the first larvae whose defences are swamped. If this is the mechanism, it would look a lot less like altruism. Let us assume though that the transformation into a brainworm is not the result of loss in an arms race. If so, one might think of trematode parasite *Dicrocoelium dendriticum* as having evolved a lottery gene. Trematodes with this gene take part in a lottery. If they lose the lottery, they become the brainworm and die. If they win, they are in a group in which another trematode transforms itself into a brain worm, and they have a much increased chance of completing their life cycle. The lottery gene has better replication prospects than its rival if either the trematode's chances of completing its life cycle without the intervention of a brain worm are very grim (for then you have little to lose by entering the lottery), or if many of your companions have also entered the lottery, thus reducing your chances of losing. Lottery genes of this kind seem quite common. For example, in times of scarcity slime molds aggregate to a single multi-celled fruiting body in which reproduction is a lottery. Cells at the tip of the body form spores, but those that form the supporting structure have lost the lottery and foregone their chances of reproduction.

What are the phenotypic effects of lottery genes? What is the adaptation in virtue of which they improve their replication prospects? Dawkins' point is that we are not compelled to nominate a vehicle as the bearer of an adaptation. Not all evolutionary processes involve vehicles. In the introduction of his 1982, he defends a pluralism not unlike that of Dugatkin and Reeve. So one could read this alternative in a modest way, arguing, with Dugatkin and Reeve, that the Wilsonian examples can be as well described from the perspective of gene's extended phenotypes; we have an adequate vehicle-free description of these examples.

Wilson and Sober, not surprisingly, are rather sceptical of pluralist responses to trait group selection. I take their scepticism to have three central elements. First, they suspect a fudging of the issues is in the air. After all, it cannot be the case *both* that Maynard Smith, Williams, and Hamilton refuted group selection *and* that group selection is really, in the end, just a version of standard Darwinian individual selection. I think this is right. We should regard broad individualism (and even more Dawkins' conception of selection without vehicles) as an important shift in our conception of evolution rather than a vindication of the 1966 status quo. It is a shift that recognises the importance of social environment and population structure in behavioural evolution.

Second, they suspect that individualist redescriptions involve an "averaging fallacy." Suppose that the co-operative dam building is evolving in a beaver population because the boost to beaver productivity in well-maintained dams outweighs the cost of free-riding in mixed groups. The averaging fallacy just averages the fitness of co-operators in all the groups in which they exist and compares them to the defectors. In our scenario, the productivity boost in co-operative ponds leaves the average co-operative beaver fitter than the average non-builder. Hence, co-operation evolves because the individual organism is fitter in virtue of being co-operative. In effect, the average fitness figures for co-operation and defection sum the results of all these selective processes to yield the result of selection. The process of averaging bleaches out all the information about process, all information about the mechanism or mechanisms in virtue of which the co-operating beaver is fitter than its rival. There is surely much justice in regarding this as a very misleading picture of the evolution of co-operation. Hence, Wilson and Sober think of "broad individual" selection as trivialising the idea that the individual organism is the vehicle of selection.

The proper conception of an evolutionary change must retain information about the process of that change, and should not just report an average result. But there is a way of understanding "broad individual" selection so that it does not merely average the fitness of every type. Let us suppose that female lionesses' tolerance of one another's cubs suckling has evolved by kin selection. On the trait group conception, the greater fitness of suckling-tolerant groups outweighs the within-group fitness advantage of free-riding. There is an alternative story here that does not commit the "averaging fallacy." A broad individualist says that the fitness of individuals in the groups with (many) altruists is higher than the fitness of the individuals in the selfish groups. On this reformulation, the broad individualist recognises that two elements combine to explain an organism's fitness: the contribution to the fitness the organism has from being in a group of a certain kind and the contribution from its role in that group. The fitness of groups drops out.

On one view, suckling tolerance evolves because suckling tolerant groups are more productive; sufficiently so to outweigh individual selection for suckling intolerance. On the other, the average suckling tolerant lioness is fitter than intolerant lionesses, because she is likely to live in suckling tolerant groups, and despite the fact that her behaviour in that group does not give her any relative advantage over her pride-mates. Intolerant lionesses' fitnesses are not depressed by the role they play in their prides. But since most live with equally mean sisters, their average fitness is depressed by the character of their social environment, despite the fact that the necessarily rare intolerant females in tolerant prides do best of all. Since both pictures recognise the importance of the division of the population structure into prides, and recognise that an organism's fitness depends both on the character of the group it inhabits and its character in that group, it is surely very tempting to agree with Dugatkin and Reeve: these pictures are equivalent. Defenders of the "extended phenotype" may exploit the same idea. So long as the replication potential of the gene depends both on its place in the meta-population of genes and its role in the local population, no averaging fallacy is committed.[11]

Third, Wilson and Sober argue that the conception of trait group evolution "also predicts that natural selection can operate on units that were never anticipated by kin selection and game theory, such as multi-generational groups founded by a few individuals . . . and even multi-species communities" (1994, p. 597). This suggestion does highlight something of profound importance, easily overlooked by those in the grip of the standard picture of kin selection: trait group population structure can be of profound evolutionary significance independently of considerations of kinship, for the fitness effects on an individual of trait group membership depend only on variance between trait groups, not on the mechanisms which cause that variance. However, recognising the contribution to an organism's fitness deriving from its place in this population structure does not require us to conceive of trait groups as vehicles.

So Wilson's and Sober's reasons for rejecting pluralism are not convincing. Moreover, pluralism mitigates a serious objection to the "trait group" analysis. The canonical examples of trait groups often do seem like high level vehicles. Ant colonies and vampire bat roosts are not theorists' inventions but parts of the biological landscape. But these examples make it easy to overlook how many populations emerge as structured into trait groups on their analysis. These include cases which do not seem like the selection of groups at all. If one were to insist that the trait group description of these cases is the only correct description, and that trait groups are high level vehicles, then these cases emerge as important counter-examples.

Trait groups are linked by their "common fate." But what is common fate? Our trematode parasites might be thought to share a common fate, in a certain Irish kind of way. Even though one dies in the ant to send her fellows to the cow, all enter the brainworm lottery and that is their common fate. But some of the examples of Wilson and Sober (1994) trade on a still more attenuated conception. To bring evolutionary games theory within the trait group ambit, they need a very inclusive sense of "common fate." Recall the water crickets; on their view, improved co-ordination evolves by selection of trait groups consisting of the boating pair, for the better co-ordinated cricket "can evolve only by causing pairs to succeed relative to other pairs" (1994, p. 596). Despite their ephemeral nature, the two crickets boating together have a common fate. So Wilson and Sober treat the pairs as groups with a fitness value even when thinking of the evolution of defection (1994, p. 597). So on their view, the pair consisting of a naive cricket and a partner that drowns him is a vehicle with low fitness. But surely there is no sense in which the defector and its victim have a common fate in virtue of their shared journey across the pool.

Other examples that make the same point. The fictional Greenbeard Gene has enjoyed a modest fame in the evolutionary literature. Carriers of the Greenbeard gene (a) have Greenbeards and (b) aid other Greenbearded organisms. Now Greenbeard genes have often been taken to be outlaws. The paradigm outlaws are meiotic drive genes which undercut the fitness of the organism they inhabit as a consequence of their biasing their prospects of reaching the gametes. Similarly, Greenbeard genes sacrifice the interests of the organism they inhabit to help copies of themselves in other organisms. Yet on Wilson's account, the Greenbeards are a trait group. The Greenbeard trait cuts out a sphere of influence: the organisms that are the potential beneficiaries of the organism carrying the "sacrifice yourself for the Greenbeards" trait. If the Greenbearded gene increases in frequency as a result of mutual aid, it does so by group selection. It is surely very weird to interpret the evolution of an outlaw gene – if Greenbeards really are outlaws – as group selection.

The outlaw status of the Greenbeard Gene is controversial. But uncontroversial examples of outlaws look like examples of trait group selection, too. Dawkins (1982) imagines a male outlaw gene. His outlaw is a gene on the Y chromosome that biases its male carriers towards his sons, and hence towards copies of the male bias gene. The imagined male feeds his daughters to his sons. By Wilson's definition, the "cannibal-Y" males in a family will form a trait group – an exotic kin group – and if the "cannibal-Y" gene evolves, it will evolve by boosting the output of the cannibal-Y group, and it will evolve despite the fact that cannibal-Y males are less fit than normal males which have both sons and daughters that survive and reproduce. This is very paradoxical. Outlaws are ultra-selfish genes. They are genes that become more frequent in virtue of evolution below, not above, the level of the organism.

So if Wilson, Sober, and other defenders of trait group selection insist that trait groups are real units of selection, we seem here to have a series of examples that argue that their conception of common fate is too weak, and interactions between trait group members too varied, for every trait group to count as a vehicle. On the other hand, if we see the trait group analysis of these cases as an often useful heuristic, but one of several, then the fact that some of these cases do not look like examples of high level groups is no longer a problem.

This idea is reinforced in other ways. Maynard Smith (1987, personal communication) denies that kin selection *requires* a population divided into groups. He invites us to imagine a forest consisting of a pure stand of a single species. If seed dispersal is limited, distance in space will vary pretty smoothly with degree of relatedness, so neighbors will be relatives. Altruism might evolve. Of course, we *can* construe this as a population divided into groups. So, for example, we can pick a set of focal individuals, and construct our groups around these, treating (say) the ten nearest neighbours of each focal individual, plus that individual, as a group. We would thus conceive of our forest as a patchwork. Within each group a defector (say a tree that hogs the light) will do better than an altruist. But still, altruist groups might be productive enough overall for the average altruist to do better than the average light hog. The problem, though, is that the patchwork is arbitrary; there are many different, equally good, ways of constructing our patchwork.

The moral is the same. We should think of the trait group conception as a good heuristic for thinking about kin selection and social evolution more generally, but we should not think of it as the only correct view of these evolutionary episodes.

3.3 Group Selection: Old and New

I think the pluralist suggestion of Dugatkin and Reeve hints at an important discontinuity between old style group selection and trait group selection. An organism, as the trait group theorists frequently remark, is a population of cells. So do we have two equivalent, equally good descriptions of the evolution of some paradigmatic adaptation of an individual organism? One would speak of the relative fitness of individual organisms. The other would identify the individual cells as vehicles, and recognise two vectors in their fitness: a fitness component from their role within the population of cells that the vulgar think of as an organism, and the fitness they derive from the character of that population vis-à-vis other populations. This generalised version of broad individualism fails to recognise the importance in evolution of the

96

organism. The idea of the vehicle recognises and identifies the organism's role, but only if the evolution of the vehicle is an objective feature of evolutionary history. The same, arguably, is true of other highly co-adapted populations: termite nests, ant colonies and the like.

This contrast in examples suggests that there is lurking ambiguity in the group selection debate. One strand of this debate is the attempt to characterise *population-structured selection*. These form the central pool of examples of those who defend trait group selection. My suggestion, paralleling Dugatkin and Reeve, is that trait group selection, broad individualism, and extended phenotype theory all give equivalent formulations of population-structured selection. Population-structured selection is a precondition for the evolution of high level vehicles, and the evolution of population adaptations. For want of a better term, I will refer to this stronger sense of high level selection as *superorganism selection*. This ambiguity is reflected in the history of the debate. Old style group selectionists (especially those whose paradigm was the social insect colony) often seem to have something like the superorganism in mind, for they emphasised the integration and cohesiveness – the organism-like characters – of their collectives.[12]

Wilson and Sober are friends of the superorganism but do not think there is a distinction *in kind* between superorganisms and lioness kin groups, lion coalitions, or water cricket pairs. Our trait group becomes more like a superorganism to the extent that fitness differences within the group disappear (1989). The clasping adaptation, on this view, pushes water cricket evolution towards the superorganism end of the continuum for it suppresses within-pair differences.[13] Their 1994 paper supplements this criterion with a second. Trait groups become more like organisms to the extent that different trait groups pick out the same creatures. On this view, there is no qualitative distinction to be drawn between water crickets, lion coalitions, and social insect nests.

Though of course there will be borderline cases, I suggest the distinctions between population-structured selection and superorganism selection are quite robust. Where we just have population-structured selection, we have "mere" trait groups. We can see these as vehicles, but there are equally adequate alternatives that do not so conceive of them. But population-structured selection occasionally leads to the evolution of superorganisms, organizations of organisms which have as good a claim to being objective vehicles as organisms themselves.

In the final section, I explore some differences between trait groups and superorganisms.

4 REAL VEHICLES

I think there are two ways in which we might distinguish trait groups from superorganisms. The first focuses on the fitness of the composition, and on whether it can vary independently of the fitness of the components. The other argues for the importance of cohesion.

4.1 Trait Group Fitness and
Organism Fitness

In the last section, I sketched the broad individualism of Dugatkin and Reeve, and showed that the broad individualist can redescribe even some social insect evolution as population-structured evolution. However, the example I discussed had a rather special feature, one which makes this redescription especially plausible. The trait group story, the extended phenotype story, and the broad individualist story all identify the same adaptation. The relevant chemical products evolve through their function in warning and defence, whether we see this as an adaptation of the nest, the ant, or the genes. Similarly, we see the clasping gene as adaptive because it prevents a cricket's partner from leaving the leaf first, whether or not we think of the cricket pair as the vehicle of selection or as part of the environment.

There are examples which suggest that the fitness of the nest is not just a simple sum of the fitness of the individual organisms within it, for it can vary independently of those organisms' fitness values. In these cases, broad individualist and trait group models identify distinct adaptive effects. So counter-factual scenarios and comparative biology can discriminate between hypotheses in which the nest is the vehicle, and those in which it is the environment. For example, in some eusocial insect species, there are genes for sororicide. In single queen colonies, when a new generation of queens hatch, the first hatched kills potential rivals. Consider the sororicide gene, and its actual and potential rivals. There are (at least) two possible mechanisms by which this gene might compete with its rivals. These mechanisms identify distinct adaptations conferring fitness benefits on different vehicles.

One idea is that this gene adapts the nest itself, by decreasing genetic variation with the nest. Hence it helps to suppress destabilizing within-nest competition. The idea is exported from Buss (1987), who demonstrated that the genetic uniformity of individual organisms is no accident.[14] There are many mechanisms that function to suppress genetically distinct cell lineages within organisms, for such lineages have the potential to subvert the integrity of the organism. Presumably, the same is true of nests: the more genetically heterogeneous a colony, the greater the potential for within-colony conflict. So selection between colonies would favour variation which resulted in greater homogeneity.

The organism level hypothesis about sister-killing sees it arising through a prisoner's dilemma. Once the sororicide gene appears it will spread, even if it were from the perspective of the nest an outlaw. It is an organism-level equivalent of a meiotic drive gene. In an environment in which your nest mates are likely to carry the sororicide gene, you will be very unfit unless you carry it too. So though it increases the genetic homogeneity of the nest, that is a side-effect of the organism-level prisoner's dilemma.

These adaptive ideas are not equivalent, for we can envisage one component of the environment changing without the other varying.[15] So consider counter-factuals in which we vary the individual consequences of sister-killing. In nests in which workers protect queens that do not strike first, the prisoner's dilemma is suspended. That is equally the case if royal eggs hatch in separate parts of the nest, so queen hatchlings can avoid their sisters. In such environments, whether or not uniformity is a nest level advantage, the prisoner's dilemma hypothesis says we should not find sister-killing.

Next consider counter-factuals in which the prisoner's dilemma still rages, but without the nest level consequences we have been considering. Think of nests in which a single queen has many suitors, or nests in which queen generations overlap. In such circumstances, sister-killing would not lead to genetic uniformity. In these counter-factuals, we consider worlds in which the trait does not have the same effect on nests as it does in our world. In these worlds, there is no between-colony selection for sister-killing, so if the "unity of the nest" hypotheses were right, these worlds would not be sister-killing worlds. Comparative biology may provide a reasonable analog of these possible worlds. For example, we could hope to confirm a "unity of the nest" hypothesis by showing that we have sister-killing *only* when it generates genetic uniformity.

Nests and hives are very plausible examples of real vehicles. So it is not surprising that there are at least some hypotheses about nest adaptation which do not translate smoothly into hypotheses about individual organism adaptation. We could see warning either as advantageous for the ant, in the right social circumstances, or as advantageous for the ant nest. But if we have the adaptive mechanism of sister-killing right, the same double vision for it is not so plausible. We cannot see it as good for the sister-killers, given the fitness contribution from their social environment, or good for the hive, while focusing on the same set of causal interactions. If it is a winner-take-all prisoner's dilemma, it is mandatory for the sister-killers irrespective of the fitness consequences for them of their more genetically uniform world. If it is a consequence of hive level selection for genetic uniformity, then the sister-killers have entered a lottery altruistically. For the bee sister lottery is unlike the brainworm's lottery, where each entrant had so much to gain and a much smaller risk. Hence, we could see the brainworm lottery both ways. But if we set aside the prisoner's dilemma, it is hard to see that the bee sisters are individually better off by entering such a fatal lottery.

Let me try to make vivid the relative independence of hive fitness and organism fitness in a slightly different way. I have been considering thought experiments to pry apart hypotheses about social insect groups from hypotheses about their members. We cannot run analogous thought experiments about our water crickets and the water cricket pair. The "fitness" of the pair just is the fitness of the individuals in the pair. Knowing the fitness consequences for the interaction for each member of the pair just is knowing the fitness of the pair. It is just a simple sum of their fitness, so we cannot envisage counter-factuals in which the two come apart. The values cannot vary independently. Knowing the fitness consequences for each individual lioness, of tolerating of failing to tolerate suckling, just is knowing the fitness consequences for the pride of that trait. To know whether suckling tolerance will evolve, we need of course to know the fitness of other prides as well. But the fitness of each pride member specifies the fitness of the pride. But if sister-killing improves the prospects for the hive's survival, and its prospects for founding new colonies, knowing the fitness consequence for each would-be queen of their deadly interaction is not knowing the fitness consequence of that interaction for the hive. As we shift from superorganisms to organisms themselves, this information gap becomes still clearer. Knowing the reproductive fate of a given cell lineage (or even all of them) will often tell you very little about the fitness of the organism.

4.2 Vehicles and Their Environment

In the eyes of Dawkins, Hull, Buss, Wilson, Sober, and others, organisms are exceptional products of evolution. For Dawkins the vehicle is a *special case* of interaction. Sometimes the adaptive phenotypic effects of genes combine together in the design of a single structure in which they reside. Then and only then do we have a vehicle. When adaptive phenotypic effects come bundled together in a single integrated entity, we have a natural segmentation of an evolutionary process. It divides into the adaptations genes code for versus the environment in virtue of which those gene effects are adaptive. There is no greater difficulty in segmenting ant nest from ant environment than there is for doing the same with the individual ant, for ant nests, too, are integrated and cohesive.[16] It is perhaps for this reason that Hull, despite the generality of his aims, defines interactors as units that interact with their environment as a cohesive whole. Interactors are structured units, units with effects on one another, effects with downstream consequences on replication propensities.

Once one moves away from the insistence on interactors as complex wholes, the robust distinction between adaptation and environment comes under threat. So, for example, we can "move the boundaries" with Dawkins' "Extended Phenotype" examples. For example, we can take the adaptive effect of the male's Bruce Effect abortion genes to be that of causing newly pregnant females to abort in an environment in which he is a potential father when she comes back into season. Alternatively, we can take it to be the production of the abortifacient pheromone. That pheromone is adaptive in an environment in which newly pregnant females lack chemical protection for such pheromones and in which he is her potential mate. So Dawkins says:

> Whichever link in the chain a geneticist chooses to regard as the "phenotype" of interest, he knows that the decision was an arbitrary one. He might have chosen an earlier stage, and he might have chosen a later one. So, a student of the genetics of the Bruce Effect could assay male pheromones. . . . Or he could look further back in the chain. . . . Or he could look later in the chain.

What is the next later link in the chain after the male pheromone? It is outside the male body. . . . He chooses for convenience to end his conceptual chain at the point where the gene causes pregnancy blockage in females. That is the phenotype gene product which he finds most easy to assay, and it is the phenotype which is of direct interest to him as a student of adaptation in nature. (1982, p. 230)

The idea of the extended phenotype subverts the idea that genes always combine to operate on the environment as a cohesive whole. We can draw a clear distinction between adaptive trait and the selective environment in virtue of which a trait is adaptive only when phenotypic effects are bundled into vehicles rather than dispersed as interaction. Kin groups and temporary coalitions do not have determinate phenotypes. They do not allow an objective distinction between their design and the environment in virtue of which they are designed. So they are not "real vehicles."

Wilson and Sober resist the idea that vehicles must be cohesive and integrated. But dispersed and uncohesive "quasi-vehicles" have neither physical nor functional boundaries. I doubt that there is a clear way of distinguishing between a quasi-vehicle and its environment. That is why we can reinterpret trait group selection as selection of individuals in a particular population structure. I see three possible lines on the boundary issue:

(a) We could attempt to defend an objective way of drawing the vehicle/environment boundary, even for unbundled effects emphasized by Dawkins, Wilson, and Sober.
(b) We might accept that when, but only when, the phenotypic power of a replicator is unified into a vehicle, there is an objective distinction between adaptation and environment.
(c) We might read the arbitrariness problem back into the cases where we have vehicles. Even when we can identify an organism, or something like an organism, there is no objective way of distinguishing between adaptation and environment.

The trait/environment distinction would be unproblematic if we could appeal to development to distinguish adaptation from environment. But it is most unlikely that the causal histories of, on the one hand, water cricket co-operation and, on the other, the environmental structures that make that co-operation adaptive divide neatly into different kinds. It would be unproblematic if we could see selection as

adapting a population to its environment. For then adaptive phenotypic effects would be evolutionarily labile, whereas the selective environment would be fixed. Relative stability would distinguish world
from design. But this picture of evolution breaks down in contexts in
which the social environment is salient in selection. For then population and environment change together.

Wilson and Sober think that the trait group selection reading of
water cricket evolution is right, and the broad individualist version
wrong. So they are committed to (a), a general objective distinction
between adaptive effects and the environment in virtue of which those
effects are adaptive. It is hard to see how to sustain that distinction. If
that is right, there really is a qualitative change in the shift from population-structured evolution to superorganism evolution. Cohesion
matters. Organisms and superorganisms are real vehicles. There is a
fairly objective description of their location in design space. Their existence and location in the biological world is stance-independent. Trait
groups that are not cohesive do not share this objective existence as
vehicles.

5 CONCLUSION

I have not travelled all the way with Wilson and his allies in their revival
of group selection. But that scepticism should not obscure the importance of their reinterpretation of social evolution. First, they are right
to emphasise the great importance of population-structured evolution.
It has a pervasive role in evolution. Moreover, while it is not compul-
sory to view the process as group selection, it is a striking fact that their
picture is often a very intuitive conception of many of these episodes.
Second, they are right, I think, to decouple population-structured evolution from kin selection. Relatedness is one mechanism – perhaps very
important, but still only one mechanism – that generates important
structure in the population. Third, they are right to insist that our conception of evolution be sensitive to the mechanism of evolution, not
just its output. Finally, they have helped identify a mechanism that can
lead to real high level vehicles.[17]

NOTES

1. "Vehicle" is Dawkins's term, "interactor" is Hull's (Dawkins 1982; Hull 1981;
1988a). They are often taken to be equivalent, but that is a mistake. Hull's

notion of an interactor is an attempt to capture a general functional role in evolution, to capture the way in which competition impacts on differential replication. Dawkins' idea of the vehicle is an attempt to characterise the distinctive evolutionary role of organisms. The central message of his most important work is that genes act adaptively on aspects of their environment outside the body of the organisms in which they reside. Replicators do not always act by constructing vehicles. In this chapter, neither term is wholly apt. For I think there is an ambiguity within the group selection debate between the general notion of interaction and the more restrictive organism-like idea of the vehicle. But as Wilson and Sober use "vehicle," I shall too.

2. D. S. Wilson's take on these issues has been developing over some years. See especially his 1983, 1989, 1990, 1993, 1997, and 1999. Sober has been a powerful ally, especially in their recent collaborations; see Wilson and Sober 1989, 1994, and Sober and Wilson 1998. The idea that kin selection is an instance of group selection has been independently developed by Colwell (1981). Hamilton accepts that Wilson's line on kin selection is at least one possible reading of his work; see Hamilton 1975.

3. Friends of Dawkins' idea of the extended phenotype are well-placed to resist this argument. He argues that we should not assume that the phenotypic effects of replicator teams come packaged as discrete co-adapted vehicles. If we could see the evolutionary history of the organism itself, there would be a stage at which important phenotypic powers of genes were expressed *neither* as adaptations of a cell, *nor* of organisms. The evolutionary history of traits central to the invention of the organism did not depend on the selection of diffuse organisms, but instead depended on selection for adaptive effects which were not features of vehicles. Similarly, there is a point in the evolutionary history of adapted groups in which genes had effects which were not expressed as traits of any vehicle. So we can think that termite colonies are vehicles, and we can doubt that their less co-adapted precursor groups are, without thinking that adaptations of groups evolve by selection on individual organisms.

4. Wilson does not restrict trait groups to organisms within the one species; one of his examples is of mixed-species flocks of birds. So the exact definition of trait groups is an issue of some delicacy. Though beaver predators are within the sphere of influence of the dam-building trait, they are not within the trait group, for a good dam impacts on their fitness in a direction opposite to its influence on the beavers. But why aren't beaver fleas in the trait group? They seem like lazy beavers in the dam: they benefit without aiding. So if the defecting beaver is in the trait group, perhaps so should be the beavers' fleas.

5. Moreover, the formal models often assumed very large within-group fitness differences and relatively weak between-group differences. For altruist groups were able to show the benefits of altruism only in the relatively exceptional circumstances of selfish groups going extinct.

6. Neander (personal communication) has objected that there is an important difference between the defence chemical and an ordinary adaptation. The defence chemical (but not, say, the camouflage stripes on a tiger) only ben-

efits the ant via benefiting the nest. It is only because of the benefit to the ant's nest that ants with defence chemicals are on average fitter. But this response seems to beg the question against the broad individualist by pre-supposing that nests are vehicles, and have fitness values in the same sense organisms do. Think of any adaptation that improves the caddis fly larvae's *physical* nest. The broad individualist thinks that the caddis fly larvae is fitter in virtue of that adaptive shift, but not because the nest is a vehicle. The nest is part of the fly's environment, but it is an aspect of the environment in that the fly constructs rather than merely adapts to. That is what the broad indi-vidualist says of the *social* nest, too. Nonetheless, I think there is something right about the idea that social insect colonies that genuinely do have fitness properties independent of those of their members, and I will return to this idea in Section 4.

7. Williams (1992) suggests this line of argument without endorsing it.

8. Dawkins (1982) sometimes runs a line like this. He recognises that his con-ception of the extended phenotype of a gene commits him to the possibility that genes express themselves through group structures. Yet in his 1982 he downplays this possibility; there are passages in which he seems to suggest the appearance of group selection derives from the fact that selective forces acting on individual organisms are sensitive to the social environment of those organisms (1982, pp. 240–1).

9. Synchronised breeding seems important in the evolution of large flour beetle groups in the Wade experiments, because it cuts down on cannibalism.

10. Broad individualism, in turn, he thinks has several equivalent formulations. Chapter 2 and Waters (1994) also defend "pluralist" interpretations of evolutionary theory, though their focus is gene versus individual organism selection, not group selection.

11. Paul Griffiths has suggested (personal communication) that this conception of an organism's fitness as having two sources may understate the complex-ity of real selection processes. We know that there can be very complex many-one relations between evolutionary shifts at the level of genes and at the level of the phenotypes those genes express. A simple phenotypic shift can conceal complex genotypic shifts. Equally, perhaps simple shifts in the nature of trait groups may depend on very complex shifts in the fitness of the component organisms; shifts whose complexity is not captured by the simple dichotomy of fitness from within group, and fitness from the nature of the group. Perhaps so, but then the same problem is an objection to the trait group model. For it too trades in only two fitness commodities: the fitness of trait groups and the fitness of individual members of those groups.

12. Though as Wilson has pointed out to me, quite often this does not seem to move beyond rhetoric. Their models have been, like the trait group theorists', models of the evolution of particular traits.

13. Perhaps not very far. For the clasping adaptation may well not be the end of the defection/co-operation arms race. One might well read Wilson and Sober as thinking that superorganisms have evolved only when fitness differences within the trait group have gone for good. We have a superorganism when the component organisms are in a fitness-equalizing evolutionary sink –

when, in Dennett's terms, their commitment to their fellows is ballistic, not up for renegotiation (Dennett 1994).

14. Genetic diversity within the one organism signals potential conflicts of interest, but that potential is mostly blocked by mechanisms that ensure that no DNA in a given organism can replicate except by aiding the replication of other DNA. Hence, the germ-line DNA has a common fate. This connects importantly to Dawkins' conception of an organism as defined by the developmental cycle. For if there were no developmental cycle, we should not expect fate to be common: The way would be open for distinct cell lineage to control different avenues of reproduction, and hence common fate would be undermined.

15. I think the suggestion here is an informal equivalent of the criterion on high level selection run by Lloyd (1988), and Damuth and Heisler (1988), who define high level selection by appealing to the irreducibility of the fitness of high level interactors to the fitness of lower level interactors.

16. Though the genotype/phenotype relation is different in superorganisms than organisms. Migration in and out of a superorganism, and deaths of component organisms and their genes, imply that genetic inheritance in superorganisms can be Lamarckian. A phenotypic change can cause a genetic change in a superorganism – a loss or gain of an important gene – which is inherited to the next superorganism.

17. Thanks to Richard Dawkins, Dan Dennett, Paul Griffiths, David Hull, Peter Godfrey-Smith, Philip Kitcher, John Maynard Smith, Karen Neander and Jim Woodward for their comments on earlier versions of chunks of this chapter. Thanks especially to David Sloan Wilson for very extensive correspondence on all the issues in this chapter.

Part III

Evolution and Macro-evolution

5

Punctuated Equilibrium and Macro-evolution

I INTRODUCTION

Hardened Darwinism is the view that the history of life is explained by forces operating on populations of organisms, or perhaps, on their genes. Large scale events and processes are no more than an aggregation of the fate of individual organisms. This story has received some rough treatment in the literature lately, for hierarchical views of evolutionary theory are currently popular. Eldredge, Gould and Stanley,[1] to name only three, are on the record in urging that there are robust macro-evolutionary phenomena, phenomena not captured by generation by generation change in gene frequency. Three large scale patterns are most on people's minds. One is mass extinction. Mass extinctions, it has been alleged, are regular and intense. Most important, surviving mass extinction is mostly chance; it is not determined by an organism's suite of adaptations. If so, the shape of the tree of life is determined not by the relative fitness of its various twigs, but by their proximity to an extrinsic, perhaps even extraterrestrial, pruner. A second is Gould's most recent preoccupation, the "cone of decreasing diversity." The theme of his 1989 is his idea that life reached its maximal diversity at the Cambrian explosion; diversity has declined since. On Gould's view, natural selection played no central role in either the establishment of that diversity or its decline.[2] But my focus is the hypothesis of "punctuated equilibrium." At a minimum, this is the claim that the fossil record reveals that a species' typical life history is rapid formation followed by stasis for the bulk of its life span, then disappearance by extinction or by splitting into daughter species. Species do not typically form by gradual transformation out of their ancestors and disappear by gradual transformation into their descendants. This pattern is alleged to have two striking consequences. In some of their early work, Gould,[3] Eldredge and Stanley have taken it to have implications about the process of change. The changes we see in the disruptions of stasis are too rapid and too large to be effected by a generation by generation shift of gene frequencies in a population. The changes are not effected by natural selection working on normal variation in a population. Species are established by something like macro-mutation. Moreover, the stability of species requires explanation. Gould has written of the "paradox of the first tier" (1983, 1985). If natural selection were the only force operating on populations, we would expect them to converge on optimum design. Since they do not, there are higher level forces in action. More recently, Eldredge in particular (but also Gould) has seen the hypothesis of punctuated equilibrium as part of a theory of biological hierarchy. If the pattern hypothesis is right, species are subjects in their own right of evolutionary laws and causal explanations. The role and fate of a species is not a mere summation of the role and fate of its members. Let me try to unpack this a little, concentrating to begin with on the different ways the punctuational hypothesis can be taken.

II PUNCTUATED EQUILIBRIUM: THREE VERSIONS

II.1 Pattern Hypotheses

We can take the punctuational hypothesis as just the claim that the typical life cycle of a species is one of quick origin, morphological stability, and rapid disappearance. So understood, its two crucial empirical claims are that the apparent gappyness of the fossil record is real, no mere artefact of a very incomplete record, and that the tempo and mode of evolution is not uniform.

If the pattern is genuine, it poses some difficult questions: (a) how is stasis maintained (b) how and why does it break down (c) what are the mechanisms of the punctuations (d) what then is the nature of species and speciation? In answering these questions, the idea runs, we must move beyond the idea that evolution is nothing but the gradual change in the composition of a population, and the associated fictionalist view that species names are of arbitrary segments of lineages that change fairly smoothly over time.

There is doubt about the evidence for the pattern. There is good reason to expect stasis to be over-estimated. Species with broad ranges and/or large populations are more likely to be exemplified in the fossil record. But they are also more likely to exhibit stability in form and behaviour. Moreover, there is a hard part bias. Almost always, only hard parts are fossilised. The transformations over time of soft parts, biochemistry, and behaviour largely escape the record. So fossil traces will preserve the variation of only a fragment of a species' traits. Moreover, since so few organisms leave fossils, the record of a species is likely to understate variation in that species.

Despite the problems of vindicating the pattern hypothesis, I am going to assume it's true, and will explore its ramifications.

II.2   Process Hypotheses[4]

Considerations about punctuated equilibria have been supposed to fracture hardened Darwinism in two ways. First, it has been supposed that change requires a saltation, a change from one species to an other without intermediate forms.[5] Second, it has been supposed that stasis, the stability of a species' form over long periods requires explanation; an explanation the neo-Darwinian cannot give, since that theory predicts not stasis but slow change.

II.2.I   CHANGE. There is a standard argument against macro-muta-tions. They are possible, but are almost certain to be unfavourable. For most points in genetic space are unviable. Viable combinations are islands in a huge sea of unviability. Any organism is on the shores of one of these islands, though perhaps not on quite the highest point. A small change might take it up the slope a bit. But a large change, unless by very rare chance it finds a new island, will doom the organism's offspring.[6] Moreover, Mayr has an alternative theory of fast change. A geographically and reproductively isolated population at the periphery of its normal range is likely to evolve quickly and without fossil trace. For such groups are small, so drift and natural selection can work more quickly. Moreover, these isolates are under strong selection pressure, for they are not in their normal niche, hence are not well adapted. Since they are isolated, changes are concentrated, and are not diluted by migration.[7]

Evolutionary change can be rapid in those circumstances, yet the intermediate forms are well adapted to neither their former niche nor the one now being colonised. So they will be neither long lasting nor widely distributed. So it is no surprise that these intermediaries are not found in the fossil record. This hardened story is so convincing that these days Gould denies that he ever had macro-mutation in mind as an account of punctuational change, a denial that sits uneasily with a good deal of his early rhetoric and of his many complimentary references to Goldschmidt.

II.2.2 STABILITY. Gould has argued that the stability of species constitutes a threat to hardened Darwinism. He thinks that morphological and developmental constraints sharply limit the power of natural selection to change a species' basic structural design. These constraints can break down only in special circumstances. Gould conceives of the hardened Darwinian as implicitly thinking of natural selection as a mighty transforming agent, slowly but inexorably shaping an organism to its niche. That view of selection suffers from a "paradox of the first tier" (Gould 1983, 1985). If natural selection were the most important force shaping the biosphere, we should see life being pushed in the direction of greater optimality. But we do not. Species usually do not change. Moreover, there is no sense in which life is better adapted now than in (say) the time of the dinosaurs. So stasis is an anomaly for those who expect natural selection to optimise organic design. So Gould urges us to see punctuated equilibrium as not just a claim about pattern, but one about process as well. Evolution involves the differential survival of species, not just their members.

I don't buy any of this. I am not convinced that stasis is an anomaly for those who think that natural selection is the only selective force operating on individuals in a population. Moreover, neither species nor any other high level selection is a suitable explanation of stasis. Let me take these ideas in turn.

There are good reasons for expecting species to be stable. Stability can be the result of habitat tracking. As the environment changes, organisms may react by tracking their old habitat. They might move north as the climate cools, rather than by evolving adaptations to the cold. Selection will usually drive tracking. For migrants that follow the habitat (personally or by reproductive dispersion) will typically be fitter than the population fragment that fails to move, for the residual fragment will be less well adapted to the new environment and will be faced with new competition from other migrants tracking their old habitat. The evolution of developmental canalisation can stabilize the phenotype, as does selection itself. As a population approaches optimum fitness for the particular fitness island on which it finds itself, deviations from that optimum are punished, even if the island is only a local optimum, and even if there is a superior genotype quite close to the local best.

Moreover, we need not invoke high level selection to explain imperfect design. We know natural selection is not the only force operating on the genetic make-up of a population; drift is another. Moreover, if the day-by-day processes on populations are calamitously disrupted every twenty-eight million years or so, whatever optimizing processes there will be disrupted too, but not because of a higher level selective force.

Furthermore, we should distinguish adaptation from optimisation. I can think that every trait is an adaptation without thinking that any are optimal. Thick fur, for example, is an adaptation for insulation if creatures with thick fur have survived longer or reproduced more fecundly in virtue of its insulating properties than some of their rivals. Nothing in this story presupposes optimality; some of the competition may have succumbed by ill luck; drift and selection might combine to explain the widespread possession of fur. Moreover, fur is an adaptation for insulation even if it is not an optimal adaptation to insulation; others may have developed both fur and fatty insulation under the skin.[8]

So even if natural selection is the only force, and even supposing it to act everywhere and always, we would not expect organisms to have, or to be approaching, perfect design. Moreover, even if life were locally optimal, we would not expect the history of life to be a history of progress. For optimal adaptation is relative both to local conditions and the constraints imposed by the species' history and engineering. Organisms in radically differing environments, or with radically differing designs, cannot be compared for goodness of design. On both counts, it makes no sense to say that (for example) sharks are better adapted than tigers. Yet the picture of life's history as a progress requires just this empty decontextualized notion of optimality. Not even a very optimistic adaptationist, therefore, predicts a long succession of ever better adapted biota. They do not expect a march of progress, so there is no need to posit an extra, high level, mechanism to disrupt the march.

Stasis may well be real; no artefact of the record, but still be a by-product of common or garden natural selection.

But imagine we do need to explain why micro-evolutionary processes fail to transform a species. The shift from thinking of punctuation as a "Goldschmittian" process to the picture of it as providing the variation for a high level selective force robs the defender of punctuated equilibrium of the resources to explain stasis. If punctuation is an event or process that somehow has the powers to crack the shell of homeostatic processes, architectural and developmental constraints that normally grip a species, then it is a candidate explanation of stasis. We explain it by explaining why those events are rare. But this is to see stasis and its breakdown as a variant process at the level of organisms in a population. Species selection cannot explain stasis since stasis and its breakdown will be part of the theory of variation on which species selection acts. Just as the rate of mutation is part of the boundary conditions of natural selection, rather than something natural selection explains (except perhaps in the very long term), stasis and its breakdown will be part of the boundary conditions of species selection. In any case, adding an extra selective force will not explain why things stay the same unless, miraculously, the extra force automatically cancels out the first.

So just as we don't need any fancy mechanism to explain change, we don't need to depart from hardened Darwinism to explain stability either.

II.3   Species as Agents

Macro-evolution does not stand or fall with fancy mechanisms. An analogy from the philosophy of psychology might make this clear. Cartesian dualists, no doubt, think that psychological processes are irreducible to neurophysiological ones. But functionalists have shown you do not have to be a dualist to think that psychology has a limited but significant autonomy. Each psychological mechanism is neurally realised; we must be able to give a neural explanation of how the mechanism works. But, the idea runs, we would lose explanatory power if we dispensed with psychological explanation in favour of neural ones. Why this is so is a matter of considerable debate. One popular idea is that a psychological mechanism can have distinct neural realisations. So dispensing with psychology would dissolve a unitary phenomenon into myriads of subcases. However this debate goes, the psychology-neuroscience relation gives us a rough model of how macro-evolutionary processes might be irreducible to changes in populations of organisms, even though they are realized by, and composed from such changes. It gives us a rough model of how macro-evolutionary explanations might be indispensable yet not require bizarre mechanisms.

I think we can extract a reasonable macro-evolutionary thesis in which the punctuational hypothesis plays a central role.[9] For it gives aid and comfort to the view that species are agents in the evolutionary story, by (i) showing that species can be identified, and by (ii) showing them to have a causal role. Consider first identification. If the pattern hypothesis is true, species' names do not merely pick out arbitrary segments of a continuum of variation; they do not come into existence by transformation out of their ancestor. Even though the process is not literally instantaneous, species speciate. They come into existence through splitting. It does not even matter if they change between formation and disappearance so long as those changes do not constitute speciation. Species on this view are identifiable segments of a lineage.

Moreover, species are important because speciation is important. Morphological transformation, the invention and establishment of new adaptations occurs at the same time, and may even be the same process, as the division of a species into daughter species. Furthermore, if lineages do not change by transformation, then long-term trends in lineages can hardly be the result of their slow transformation. But there are many examples of such trends. Famous examples are size increase in the horse lineage, and brain size increase in the homo lineage. So perhaps these trends are caused by differential extinctions of species in the lineage. Horse species of various sizes have been born, but the larger species have survived better.

III SPECIES SELECTION

We can see the theory of punctuated equilibria as a plausible revision of orthodoxy if we see it as showing that species are agents. It would thus be an argument for seeing a hierarchical structure in evolutionary history. One way of emphasising the significance of species is to argue that evolutionary history is in part the story of the differential selection of species, not just their members. But we must at least distinguish between species sorting and species selection.[10] Species sorting is the mere differential survival of species with some particular characteristic.

No claim about the causal import of that characteristic need be made. Everyone agrees that species are sorted; that, for example, species with the distinctive features of trilobites have failed to survive. No-one, as Philip Kitcher has reminded me, has ever denied the existence of macro-evolutionary effects; the radical suggestion is that there are macro-evolutionary causes. Species selection is a much more robust claim; a species' fate is affected by some property of the species itself, not just the properties of the organisms that compose it. But even this characterisation of the issues is too crude.

David Hull and Richard Dawkins have both pointed out that talking of genes, organisms, or species as selective agents is ambiguous.[11] There are two components to evolution under selection: replication and inter-action. For there to be evolution there must be sequences of replica-tors, of entities that are copied and, through being copied, pass on information-bearing structure to the next member of the sequence. Successful replicators in this way root long and bushy lineages. On both Hull's and Dawkins' view, evolution is the differential growth of lin-eages. But lineages grow, or fail to, in part in virtue of their interaction with the environment. The elements from which they are formed need to maintain their structural integrity, gather resources, and replicate through interaction with their environment. In some evolutionary re-gimes, the two components of the evolutionary process are matched by a specialization of entity to role. There is not just replication and interaction, but also replicators and interactors, or vehicles, though this specialization of function cannot be complete; replicators must inter-act, though interactors need not replicate. So Dawkins, for example, argues that though organisms are vehicles, and in that sense are units of selection, they do not replicate.

So, in confronting the suggestion that species are units of selection, we need to consider whether they are units of replication or interac-tion or whether, without specialisation, they do both. Orthodoxy is clearly under threat if species are interactors; it may be stretched even if species are active replicators.

IV   DO SPECIES REPLICATE?

I do not think it should be very controversial that we can see species as replicators. It is much less obvious that we must so see them. The significant structure passed on in speciation is not range or habitat, but the information in the species gene pool.[12] It is true that an isolated fragment of a population undergoing speciation will be less genetically diverse than the parent pool, and will be in some way modified by selection. So species are rather crude replicators, passing information across a speciation event with some but not drastic modification. It does not follow that from this alone that we need to see species as replicators. Some of the actual and possible examples of the recent literature are, I think, best read as exploring the idea that it is essential to see species as replicators. Maynard Smith and Sober,[13] for example, give potential examples of "species selection," in which the causally relevant property is a trait of the individuals in the species. Altruistic behaviour is often considered a candidate for explanation by high level selection; so too is the prevalence of sexual reproduction. Sober considers the effect of individual dispersal capacities on the propensity to speciate. In all these cases, the traits in question are those of individual organisms. The diagnostic question, in filtering out replicator selection, is "Who benefits?" In Maynard Smith's and Sober's examples, the benefit falls to a species, and we benefit in seeing a species as a replicator, a replicator whose replication potential is increased if the individual organisms composing it have certain characteristics. What is to be explained is the surprising prevalence of species composed of organisms that reproduce sexually, or the existence of species whose members behave altruistically. Seeing species as replicators is mandatory, it has real explanatory bite, if the prevalence of sex or existence of altruism has no other explanation. We are then required to expand our picture of the evolutionary process: not just genes and genomes, but also species are amongst the undoubted replicators. But if Kitcher and I were right in Chapter 2, there can be alternative good pictures of a selective process. So one might see species as replicators even if sex, altruism, or dispersal characteristics can be otherwise explained. For that reason the idea that species are replicators just in itself may not revise the orthodox tale of evolution.

V DO SPECIES INTERACT?

We move beyond the synthesis if we see species not just as replicators but also as vehicles. For then they are not just the scoreboard but players in the game. If the diagnostic question in detecting replicator selection is "Who benefits?," then in detecting vehicle selection it is

"Who acts?" If they act, who do they act for? We can safely assume that their speed of reaction will be slow; they can scarcely be seen as vehicles of particular genes. If they are interactors, the replicator they benefit is the species as a whole, or some substantial fragment of it. So if species are vehicles, there must as well be high level replicators. For a species to act, three requirements must be met: (i) they must be biologically coherent entities. They must be located in time, space and ecological context. They must have some internal organization; (ii) a species must stand to its gene pool in something like the way an organism stands to its genome. Species must have something like a phenotype: It must have characteristics through which it interacts with the world in order to assist its own copying or that of some other replicator; and (iii) we need an account of the empirical and conceptual distinction between, for example, a species going extinct because of species selection in favour of large-brain hominids, and a species going extinct because its small brain members are outcompeted by the members of a larger brained species. To draw this distinction, there must be phenomena explicable only by appeal to species-level causal processes.

VI ARE SPECIES COHERENT?

If species are vehicles they must be biologically coherent.[14] They must have temporal and physical bounds. Extinction imposes a reasonably determinate end. If speciation is splitting from an ancestral species then species have reasonably distinct beginnings too. Eldredge points out that species have distribution limits which impose a spatial bound analogous to an organism's skin. It is less clear that they are ecologically localised. It is common to speak of species being adapted to particular ecological niches, but it is typically false. The communities into which a species is divided often have quite different life habits and find their way into different environments. Think, for example, of differing populations of deep bush and suburban adapted possums in Australia, or the differences between Australian and feral New Zealand possums. If a species of possum has found its way into differing niches, there may be no selection force operating on the possum species as such. This fact is not fatal to the idea of high level vehicles. Selection for low population density might preserve one species of desert dwelling numbat, and drive a more densely congregating one to extinction, even though not

118

all numbats of that surviving species were desert dwellers. So Eldredge now thinks of avatars, a breeding fragment of a species in a niche, as his candidate for a supra-individual unit of selection. By this he had better mean that avatars are vehicles acting to enhance the replication of the species from which they are drawn, for avatars are even cruder replicators than species. I have mentioned that species are only replicators in a rather rough sense. The gene pool of the daughter species is distinct from that of the parent through both drift and through selected changes in genetic composition. Avatars are replicators in still rougher a sense, for the next generation is unlike the current one not just through selection and drift but also immigration. Avatars, unlike speciating fragments, are not ex officio isolated from other populations. But avatars, perhaps unlike species, are physically, temporally and ecologically localised enough to be agents.

But have avatars or species any structure, any internal organisation? No-one would pretend that they have the complex internal organisation characteristic of complex multi-celled animals, but Hull, for example, warns us against this zooaphilia. Metazoans are far from the typical organism. Still, if species (or avatars) act, there must be some glue that holds them together, something in virtue of which the organisms that make up a species are part of a whole. Eldredge, in his 1985a, argues that the "reproductive plexus" binds members of a species together. But this idea faces many problems. If we rely on actual genealogical history, individuals near a speciation event will be closer in descent line to organisms that are not species mates than to some conspecifics. It is difficult to apply the intuitive notion of a species as a reproductive community.[15] Some individuals that could interbreed do not. We have special problems applying the notion of a reproductive community to geographically scattered species divided up into races, to temporally scattered fragments of a species, and to species whose members have strong mating preferences. Only the first two of these difficulties are eased by refocusing on avatars rather than species. The notion of a reproductive plexus is in danger of covert circularity, relying on a notion of the species in deciding the real limits of the reproductive community. In recent work, Eldredge has tended to shift from Mayr's original idea to the idea that species are defined by a shared "Specific Mate Recognition System." But this seems to either recapitulate the same problems of bound, or, if these recognition systems are anatomically defined, we revert to the notion of specieshood defined by shared anatomical characters that Eldredge was in flight from. We revert to the notion of species as a class.

So species and avatars may have sufficient coherence to count as vehicles aiding the copying of high level replicators. I am much more doubtful about finding their phenotypes.

VII DO SPECIES HAVE PHENOTYPES?

If species are vehicles, they must have characteristics that aid replication. They must be adapted to aid replication. What might this involve?

Firstly, the phenotype must be built from irreducible species properties, not properties of its members. But there is no clean criterion of a species-level property, though plausible examples include niche breadth and geographic range. These are not just species-level, or at least population level, properties, they are biologically important. Generalist species, and wide ranging species, are less apt to go extinct, and are less apt to speciate. But though range, for example, is not the property of any individual, it may be a simple and direct consequence of some individual morphological or behavioural trait. Consider the groups downstream from a splitting that vary in range. If that variance is the result of an individual difference (e.g., a nesting preference, or an anti-predator strategy), it seems reasonable to regard range as reducible in an interesting way to individual traits. The failure of one descendant population, and the success of the other, would not demonstrate that species are vehicles.

Second, the trait must be selectively relevant. There are irreducible properties, for example the total number of individuals in the history of the species. We would like to know if there are any that are causally salient to selection. But causal salience isn't enough. Individual properties can be relevant to the process of selection without being adaptations; for instance, the imperfection of genetic replication. If there were no mutation there would be less variation. But the propensity of genes to very occasionally mutate is not an adaptation; it's not something that is selected for. Similarly, population properties can be relevant to the process of selection without being adaptations. A lineage may be apt to split, for the individual organisms that compose it may be spread over, and isolated in, a large number of peripheral environments. The occupation of a large number of marginal environments may well yield a strong tendency for the gene pool to split. If so, the range a species occupies is a selectively relevant property of the species, but not an adaptation. It doesn't arise because that lineage out-competes other lineages; indeed, the retreat to the margins is more likely to be a symptom of failure than success. Rather, it's the source of variation. It, like the underlying rate of mutation, is certainly relevant to selection but isn't its creature. Species interaction requires species adaptations.[16]

Moreover, added to the difficulties of reduction there is a problem about heritability. Let's consider, for instance, population density. Suppose an ancestral species splits into four daughters:

A x
B
C xxx
D xxx

Let's suppose C and D have higher population densities than A and B, and that they crash to extinction in a bad year, or that their denser population renders them vulnerable to disease. So they go extinct in virtue of a property of the population structure of the species. Suppose A is too thinly distributed over its range; it's just too hard to find a mate, and it expires too. Only the species of intermediate density B survives, and survives because of its population density. Prima facie, then, its population density is a species adaptation, and the species is genuinely an interactor. But reduction remains a problem. Once we have abandoned simple-minded pictures of reduction, population density is at least a candidate for reduction. Now, post splitting, the different population densities of A–D might well be due to a vast motley of interaction effects: there may be no salient individual properties of B that make the difference between B and the others. But in that case we have a problem about the mechanism of population regulation. I don't see how B could regulate its population (as it must if there is to be selection for a certain density) without there being a salient individual property which is the mechanism of that regulation. The demand that the property be irreducible cuts across the demand that the property be stable across generations and heritable by daughter species, characteristics a property must have to be an adaptation.

To see this, consider an example from a different context. In his [1982], Dawkins argues that we can profitably see natural selection as acting not on individual wasps, but on wasp strategies: digging a nesting tunnel versus invading an already built one. On plausible payoff assumptions, there is a mix of strategies at equilibrium: a digging frequency at which digging has the same average payoff as invading. At this point, it doesn't matter what the individual wasp does. In particular, there is no reason for it to dig at the equilibrium frequency, P^*. It can always dig, or always invade, without average selective penalty. So we can see selection as operating in a frequency dependent way to stabilise the digging-invading proportion at around P^*. The problem, analogous to population regulation, is one of mechanism. How is natural selection to stabilise that frequency except by building pure diggers/invaders at a certain ratio (analogous to a sex ratio) or by building wasps that dig at roughly P^*, and invade at $1-P^*$. So we can and should see selection operating on the wasps, not on strategies. I think this problem lies in wait for any putative species adaptation.

So it's hard to see how a species could have a phenotype. First, as we have just seen, it's hard for species to have traits that are both heritable and irreducible. Second, as Maynard Smith and others have argued, the subversion of high level processes by more powerful lower level ones makes the evolution of species adaptations for species replication most difficult. The generation time of individual organisms is swift compared to that of species or avatars. Sexually reproducing organisms exchange genetic material, whereas animal species rarely do. Moreover, there are many more organisms than species, so selection has more variation with which to work. So we can expect selection on organisms to be faster and more powerful than selection at higher levels, and to undercut high level adaptation building when the two forces are in conflict. Third, as Dawkins has pointed out, the development of complex adaptation is tied to a developmental bottleneck, the growth of the new organism from a single cell. This developmental bottleneck ensures that organism level selection is not subverted at the cellular level, for the replication fate of all the cell lineages funnels through the single germline. But still more important, Dawkins argues[17] that the initiation of a new developmental cycle is essential to the development of adaptation. A favourable change, by acting early in development, can change the whole organism, and can thus initiate the construction of something new. Reproduction "sets the switches back to zero" by beginning from a single cell. Literal beginning again from one cell is surely not necessary; there are intermediate cases like the caterpillar/butterfly transition. But species' adaptation building requires the new population to be especially plastic, if the species phenotype (not the phenotype of its individual members) is being rebuilt from near scratch. Gould and others have sometimes written that the forces maintaining stability break down at speciation events, but of course that plasticity is plasticity of organism phenotype, not species phenotype.

I do not think I have shown that we can rule out species adaptation, but it is certainly not easy to see how they could arise.

VIII WHAT MIGHT SPECIES SELECTION EXPLAIN?

On the face of it, it is going to be hard to find examples of vehicle selection that resist redescription in individual terms. Certainly, directional changes in lineages do not look good candidates. Horses got bigger; hominids got bigger brained, relatively as well as absolutely. But overall size and brain size look to be paradigmatically properties of organisms, not species. So if there has been selection for these properties, the vehicles are organisms not species. That is true, surely, even if small horses found life tough not because of larger members of their own species, but because of larger horses from sibling species. If the directional pattern is the result of interspecific competition, that no more licenses us to talk of species selection than the human extermination of thylacines enables us to talk of species selection against marsupials.[18] If we are to find species selection, it will be on properties like population structure, niche breadth, or more recondite properties still.

It is hard to establish that an appeal to more recondite species level properties buys us explanatory power. In a recent paper, Schull[19] argues that species store information of such complexity, and use it so adaptively, that they are intelligent. Species are not just individuals, some are smart individuals. Now, if species are intelligent, different lineages will differ in intelligence. That's a difference on which species selection would act. We can expect intelligent lineages to be more diverse, or longer lived than their less bright rivals.

Schull argues that a species' intelligence consists in its capacity to respond to change, a capacity that he takes to be surprisingly well developed. He claims:

1. Species learn; we can see a species climbing a hill on an adaptive landscape as it is acquiring and using information about its niche. Indeed we can see an organism's suite of adaptations to its niche as

rich information about that niche. This information is a property of the species, not the individual organism. For the information held by any one organism is only a subset of the information in the whole gene pool. The species learns by modifying the information in the pool under the pressure of natural selection.

2. Hill climbing is a simple form of learning, but some species do better. Feral animals re-evolve wild traits quickly, for the wild traits are genetically masked, rather than eliminated, by domestic breeding. In these circumstances, a species (more exactly, an avatar) shows the benefit of previous experience. Just as sophisticated organisms will relearn a behaviour (or a slight modification of it) faster than they originally learned it, so the feral population learns much faster because it is unmasking memories, not creating new adaptations.

3. Species can be polymorphic, exhibiting distinct behaviours or morphologies in distinct environments. Aphids, for example, switch from asexual to sexual reproduction at the end of summer. Evolutionary biologists think of this as phenotypic plasticity. Schull urges us to see it as a species' capacity to follow conditional strategies: in circumstances A, do ___; in B, do ___. This capacity in an organism, he reminds us, is a mark of some intelligence.

4. Schull buys into Wright's "shifting balance" theory of population structure to explain how species might escape local maxima; this too strikes him as a manifestation of a species' intelligence. For how do organisms gain insight; how do they come to rational expectations about what will happen in novel circumstances? They try trial and error in a simulated environment, thus avoiding the costs of error in their real environment by paying the costs of representing it.

Schull argues that the division of a species into many demes equally allows a species to use trial and error without paying the full cost of error.

I think, perhaps, Schull is right in arguing that we can see species as individuals that process information. But I accept Lloyd Morgan's dictum that we should attribute intelligence only when no more parsimonious idea serves. I am unconvinced that we get explanatory mileage out of seeing species as intelligent. Consider just one of his examples: an information processing account of the anglerfish lineage. We can, sure enough, see the anglerfish lure as carrying information about the environment and prey of anglerfish. But what can we thus explain that is not equally well explained by an account of the natural selection of the individual fish or their genes? Nothing that I can see. Schull does not show that the attribution of information processing capacities to the species itself explains anything otherwise inexplicable. I suspect that it buys us nothing.

IX   SPECIATION AS CAUSE AND EFFECT

In view of what has just been said, I am sceptical about the prospects of getting much mileage out of species selection. The fate of a species does seem to me to be an aggregation of the fate of its members; species selection buys us nothing. But I am inclined to think that speciation is a macro-evolutionary phenomenon.

In two senses, speciation is irreducible to the change in the spread of individuals' traits in a population. First, and least dramatically, speciation is, I take it, multiply realised by changes at the individual level. By this I do not just mean that the details of the formation of a new species of horse differ from those of a new species of possum. Even when we abstract away from the details of horse and possum mor-phology, variation in the speciating process remains. For example, the isolation of the speciating fragment can come about in many ways: geo-graphic separation, a shift in mating season or timing of daily activity. Similarly, if, as Eldredge supposes, speciation is a change in the species mate recognition system, those too can change in a multitude of ways. Speciation, then, is multiply realised by individual level changes in body or behaviour. Those changes share nothing interesting at that level. But, more strongly, unless speciation just is the accumulation of a modest chunk of phenotypic change, it is invisible at the individual level. Inspection of the change in individual properties of organisms in a population over time won't tell you whether speciation has occurred. If you followed the lines of descent from a family of birds down the generations, you could tell a lot about the evolutionary change in that mini-lineage; about the changes in its struggles with predators and prey, about clutch size and plumage; about the size of range and territory. But you could not tell whether or where speciation had taken place; you could not tell where descendants had become a different species to their ancestors. That is particularly so if one buys into the hairy-chested version of cladism that regards all species downstream from a split as new species even if one daughter has identical traits to the ancestral species. In that case our family of birds can become members of a new species because, by great luck, a small flock blown out to sea three thousand miles away ended up on an island and diverged before going extinct.[20] It is in this sense that speciation is invisible at the level of individual history.

But is speciation no more than a macro-evolutionary effect, an epiphenomenon of changes at the level of individuals in a population? Eldredge has argued that speciation has consequences for the course of evolution, an argument which in turn develops an idea of Mayr. He argues that speciation establishes barriers to the blending out of complex adaptation.

Eldredge has argued that adaptation is dependent on speciation. Individual selection does not proceed isolated from the processes through which species are born and die. Eldredge wants to rebut the idea that adaptive "economic" change[21] in the characteristics of individual organisms produces speciation. Some have taken such changes to produce speciation definitionally, supposing that the accrual of enough change is speciation. Others have supposed that economic change produces reproductive isolation, as changed-unchanged crosses cease to be viable, or as members of the diverging populations cease to regard one another as mates. Eldredge stands this view on its head: speciation consolidates "economic" change. The argument goes:

P1. Economic adaptation occurs initially in demes.
P2. Adaptations are not typically the result of a single genetic change, but of clusters of small genetic changes which must therefore be concentrated in particular individuals for the adaptation to be expressed in the phenotype.

but

P3. Demes are ephemeral

so

P4. Unless the change is somehow fixed, the death of the members of a defunct deme, or their integration into a new one, will lose the adaptation by backcrossing.

hence

C. Speciation is essential to preserving economic change, by preventing migration/reintegration into the larger unadapted population.

In a slogan: No adaptation without speciation.

If this argument is right speciation is neither reducible nor epiphenomenal. It is a modest, but genuine, macro-evolutionary cause. Not in any very exciting way; after all, every instance of speciation is identical to a change in the profile of individual organisms in particular populations. But a list of such changes would not suffice to explain what speciation was, or what its importance is. So there may be a modest place for macro-evolution in evolutionary biology.[22]

NOTES

1. See for example Eldredge (1985a); Eldredge (1985b), and Eldredge (1989). See also Vrba and Eldredge (1984); Gould (1980d); Gould (1983); Gould (1985); Salthe (1985); Stanley (1981). For very sceptical views of the whole punctuational enterprise, see Dawkins (1986) and Hoffman (1989).
2. I discuss Gould on diversity in Sterelny (1991).
3. Gould especially; see for example his (1980b), and (1980c) both in Gould (1980a).
4. Sober (1984) Section 9.4 and Hoffman (1989, Chapter 7) have also made this crucial distinction between pattern and process in punctuated thought.
5. Mayr seems to defend an intermediate position; the genetics of speciation is abnormal not because of macro-mutation but because founder populations are not just under greater selectional pressure, but also because they exhibit greater variability (cf. Mayr (1988a), pp. 473–4).
6. For a pellucid exposition of this, see Dawkins (1986), pp. 66–74, but the general point is uncontroversial.
7. See for example Mayr (1988a), Chapter 21.
8. For a more detailed and careful reconciliation of adaptive hypotheses with due modesty about the optimising power of natural selection see Godfrey-Smith (1996). Sober does not agree with this attempt to drive a wedge between optimality theory and adaptation; for his views on these issues see his (1987).
9. This line of thought is particularly prominent in Eldredge; see his 1985 a&b, and his 1989.
10. For the general distinction between the causally agnostic "selection of" some trait versus the causally committed "selection for" it, see Sober 1984, 3.2. For its application to species, see Vrba (1984c).
11. Hull (1988b) and Dawkins (1982).
12. For ease of exposition I will take it that our candidate replicator is the species, not the species gene pool, for nothing I say depends on this distinction. For similar reasons, in this chapter I will be neutral on the gene/organism debate.
13. For example, in Part 4 of Maynard Smith (1989), and 9.3/9.4 of Sober 1984.
14. Some see these claims as tantamount to the idea that species are individuals;

see Ghiselin (1974), Hull (1978), and especially Hull (1987). For arguments that we can pose the biological question independently of this issue see Kitcher (1989).

15. See Kitcher 1989 on the idealizations implicit in the "biological species concept."
16. See for example Williams (1966), especially pp. 96–8.
17. In the final chapter of his 1982.
18. See Sober (1984), pp. 259–62 for a more detailed discussion of a parallel example drawing a similar moral.
19. Schull (1990).
20. See Ridley (1989) for a defence of this picture of species and speciation.
21. That is, changes in survivability and resource use, as contrasted with the sort of changes produced by "sexual selection."
22. Thanks to John Maynard Smith for his comments on a very early version of this chapter, and to Philip Kitcher, David Hull, and, especially, Elliot Sober for comments on the most recent version.

6

Explanatory Pluralism in Evolutionary Biology

1 EXPLANATORY PLURALISM

In "The Return of the Gene," Philip Kitcher and I argued for an ecumenical conception of evolutionary theory. I hope to extend this message to adaptationist strategies in evolutionary biology, and to macro-evolution. In the social and natural sciences we are often confronted with explanatory independence co-existing with ontological dependence. A square peg fails to fit into a round hole in virtue of the geometric properties of peg and hole, and because of the rigidity of the peg and of the base. These properties are explanatorily salient, notwithstanding the fact that peg and base are composed from a swarm of fundamental particles none of which are rigid, and whose geometric properties are very odd. We explain the actions of agents by appeal to their psychological characteristics despite endorsing a physicalist theory of the mind. In the social sciences, we may explain a rise in crime rates by appeal to general social facts – an increase in unemployment, tension between ethnic communities, a collapse of family structure – while being perfectly clear that crime rates just are aggregations of behaviours undertaken for those agents' own reasons.

An ontological thesis is uncontroversial. High level entities are composed of and are nothing more than their lower level components. This is sometimes expressed as a supervenience claim: There can be no variation in the characteristics of the composite entities without variation in their components. There can be no change in the geometry or rigidity of the peg, without change in the particles and forces which make it up. A causal thesis is near enough uncontroversial too. The causal properties of the composites depend on the causal properties of their components.

These ontological and causal claims are uncontroversial in evolutionary biology too. No-one supposes that composites – species, clades, and groups – are anything more than the organisms that compose them. Speciation and extinction are realised by the life histories of individual organisms. The causally relevant properties of species, groups, or lineages are consequent on the characteristics of their members. Superorganisms, if there are any, are built from standard organisms.

But there is less sympathy in evolutionary biology to an ecumenical conception of explanation. For it is widely agreed that theories of high level selection must distinguish between mere group or species sorting and genuine selection of supra-individual biological entities. This idea has led to the view that it is difficult to establish the presence of "high level" selection. Very particular conditions need to be met before we should appeal to macro-evolutionary causes. In this chapter, I argue for less demanding conditions on high level selective explanations while preserving a distinction between macro-evolutionary causation, and macro-evolutionary patterns that are mere effects of lower level processes.

In pursuing that thought we need to examine the basis of, and advantages in, explanatory pluralism. How can a domain be ontologically dependent on another while being explanatorily autonomous? It is one thing to recognise the irreducibility of intentional psychology to neurophysiology; another to understand it. Moreover, this autonomy surely has limits. Most obviously, the causal character of the components constrain the processes to which we can legitimately appeal. But as well, lower level explanations sometimes "drive out" high level ones. For example, the demonstration that Clever Hans responded to small changes in body posture of his owner undercut the claim that he really could count. Robins that restrict their clutch size see more of their fledglings survive the winter. Perhaps this fact drives out the hypothesis that they restrict their clutch size so that the population of which they are a member does not exhaust its resource base.

So what is the basis for, and limits on, explanatory pluralism? Sometimes the relationship between domains is not **transparent**. A crime wave is an aggregation of individual behaviours. But it is an extraordinarily complex aggregation. The psychological profile of individual criminals shows great variance, and many within that group will show far greater affinities with many outside it than they show to some within it. There is no "average criminal" around whom this variation can be seen as mere noise, to be idealised away. This intuition is captured within the philosophy of psychology by the notion of "multiple realisation." A psychological trait – fear of spiders – is composed from neural states, but not simply. Arachnaphobes may have quite different neural structures in virtue of which they fear spiders. The behavioural and other characteristics of arachnaphobes are tracked by a high level characteristic but by no *single* low level characteristic. Similarly, the increase in crime is tracked not by an increase in frequency in some specific intentional profile, but by certain general characteristics of social organisation.

In my previous work I have had this picture in mind in defending explanatory pluralism while embracing ontological monism (Sterelny 1990a, Chapters 2 and 3). Philip Pettit and Frank Jackson (converging on the same idea as Garfinkel) have an interestingly different slant on these cases (see, for example, their 1992). They distinguish between **robust-process** and **actual-sequence** explanations.[1] Distinct explanations of the same event can both be important, for they can convey distinct breeds of modal information. Consider two explanations of the beginning of World War I. One details Gavril Princip's behaviour and its consequences. The other, the system of armed alliances characterising Europe at the time. The explanation appealing to Princip's marksmanship is an actual-sequence explanation, for it identifies *the particular possible world* that we inhabit. But if it is true that in Europe at the time, World War I was a war waiting to happen, then we could know the precise sequence of comings and goings in chancelleries and staff headquarters without knowing something very important. Namely, if the war had not started that way, it very likely would have started another. A robust-process explanation *compares our world to others*. It identifies an important class of counter-factuals. It identifies the feature characteristic of World War I worlds. Individual development might provide a biological parallel. We could know the precise sequence of the development of some aspect of an organism's phenotype without knowing something very important: whether that sequence is canalised, or in some other way buffered so that different paths give functionally equivalent outcomes.

Multiple realisation and robustness are obviously close relatives. Indeed, one might think they are notational variations of one another. For it seems that one might explain the relative autonomy of psychology in either language. We have already seen how the multiple realisation story runs; there is another that appeals to robustness. The property common and peculiar to all those worlds in which Phobic Joe avoids spiders is no neural state. The avoidance worlds are neurally varied. Rather, there is a psychological state of Joe – a behavioural programme resistant to new information – in each of these worlds, and it is this that explains Joe's sometimes odd, sometimes normal, behaviour. However, it turns out that we cannot always easily translate between these conceptions of explanatory autonomy. I shall argue that the Pettit/Jackson analysis gives us the best account of the relationship between selection and other evolutionary factors in adaptive explanation. Selection does not stand to history, development, and drift as composition to composite, so the idea of multiple realisation gets no handle here. But I shall argue that multiple realisation does yield an account of the relative autonomy of some forms of macro-evolutionary explanation.

2 ADAPTATIONISM AND ROBUSTNESS

Elliott Sober follows Lewontin, Gould, and others in portraying adaptationist ideas as optimality hypotheses (1987 and 1993). He distinguishes three ideas about the role of selection in the evolution of some character trait:

U: Natural selection played some role in T's evolution in the lineage leading to X.

I: Natural selection was an important cause of T's evolution in the lineage leading to X.

O: Natural selection is the only important cause of T's evolution in the lineage leading to X.

Sober counts O as a characteristically adaptationist hypothesis about T. If O is true, the population size was large enough so drift and other chance events were unimportant; genetic variability was such that many alternatives to T were exposed to selection; phylogenetic and developmental constraints were not so severe that alternatives to T were impossible to build; superior alternatives to T were not blocked by "trajectory problems" and so on. In this sense, T is optimal for X (1993, pp. 121–2).

I think this is the wrong conception of adaptationism. First, I think adaptationists are really more interested in function than in optimality. This is not just being picky. In their (in)famous paper, Gould and Lewontin chastise the adaptationist through the concept of a spandrel. A spandrel is a trait whose existence is not explained by natural selection (though some inessential elaborations of it may be). Thus Gould and Lewontin criticise participants in the synthesis for neglecting the possibility that striking traits of an organism may be spandrels; that is, Gould and Lewontin claim they ignore alternatives to I and even U.

More significantly, I think the notion of optimality and adaptation embodies a significant ambiguity. There is a tension in evolutionary thinking about adaptation and natural selection. On the one hand, the existence of adaptation is seen as posing the central explanatory problem for biology, and natural selection is seen as the only possible natural explanation of adaptation (Dawkins 1986). On the other, adaptation is collapsed onto expected fitness, and the expected fitness of an organism is its reproductive propensity under natural selection. Sober's official definition of optimality derives from this second move: optimality is the result of the undisturbed play of selective forces. We should distinguish between what selection undisturbed would produce and good design. The Paley-Darwin notion of adaptation is a notion of fit between organism and environment. For only then can we conceive of natural selection as an *explanation* of adaptation. Natural selection is no explanation of adaptation, if adaptation is by definition whatever selection undisturbed produces. This ambiguity leads to my most serious reservation about Sober's formulation of adaptationism. I think his conception misrepresents the role adaptationists attribute to selection, and its relations to other factors in evolution.

Let me begin with the relationship between optimality and function. Sober identifies adaptationism as the hypothesis that O (or, perhaps, I) is true of most significant characteristics of most populations. This gets something right about the idea behind adaptationist reasoning. An adaptationist intuition is that we can "idealise away" from genetic and proximal details in explaining the origin or persistence of traits. But it also misidentifies adaptationist reasoning, too. Most obviously, I think it is a stronger thesis than any real adaptationist needs to run (see also Godfrey-Smith 1991, Chapter 3). Adaptationist reasoning is functional reasoning. Consider Diamond's entertaining example of the chimp's large testicles. He defends a striking adaptationist hypothesis: their impressive size derives from sperm competition amongst chimps. Now, for large testicles to have that function in the population of chimps, does drift, genetic variability, and so forth have to be negligible? I see no reason to think so.

Minor revisions of Sober's interpretation might simply count large testicles as having a function if they were selectively superior to most, or the most common, alternatives in ancestral chimp populations. But even that concedes too much. If selection for sperm competition was *necessary* for the evolution of large testicles, that seems enough for large testicles to have the function of improving sperms' chances with respect to those from other chimps.[2] Chimps have large testicles as a result of sperm competition. That is not the complete causal history of the trait, but in causal explanation we neither can have nor need the whole truth. Edited highlights suffice. We need only modest claims about selection to underwrite the functional hypotheses distinctive of adaptationism (for more on this line of thought, see Horan 1989). While the inspection of adaptationist texts reveal the ubiquity of functional claims, they reveal no such ubiquity of optimality hypotheses.

Let's grant Sober's link between adaptation and optimality. More significant, in my view, is the tension between defining optimality by appealing to selection, and explaining optimality by appeal to selection. There are two notions of optimality in play in this literature. We have to disentangle these notions to bring into focus the best adaptationist view of the role of selection in evolution. Sober's definition of optimality turns on the undisturbed play of selective forces (see especially Sober 1987). Testicle size is optimal in chimps if selection is the only effective force operating in the evolution of testicle size in that lineage. Such chimps have Sober-optimal testicles. This notion of optimality is important. The contrast between local and global optima, for example, fits this conception well. Organisms are at a local optimum when selection acting alone will tend to damp down any small change in their phenotypes. However, the notion of good design[3] has to be defined independently of selection, if selection explains it. Moreover, this idea plays a major heuristic role in reasoning about selection, evolution and constraint. The optimal organism, in that sense, is the well-designed organism: one showing "observed conformity to *a priori* design specifications" (Williams 1992, p. 40). For example, Dawkins, Maynard-Smith, and others defend a method of successive approximation. Identify the behaviour to be expected from an optimally designed organism in some given niche, and use the variation of actual from expected behaviour to identify the temporal, phylogenetic and developmental constraints in play (see, for example, Maynard Smith 1987). Here well-adaptedness for some function or functions is the notion in play; perhaps more generally some notion of "fit" between organism and niche. The notion of good design used by Cosmides, Tooby, Symons, and others attempting to reconstruct cognitive design is clearly a notion of this type, not one committed to the undisturbed play of selective forces (see their papers in Barkow, Cosmides, and Tooby 1992).

The two notions of optimality are not equivalent. First, selection can reduce adaptedness. Gould and Lewontin themselves note that selection can drive a change through a population which leaves all members less well-adapted. A gene which doubles the clutch size would sweep through a population of fixed size. At the end, absolute expected fitness has not improved; indeed the metabolic and other costs of increased egg-laying might cause an eventual slide to extinction. Nonetheless, the relative fitness of fecund layers will be greater than that of modest layers, and that relative difference is critical to selection. The invasion of a selfish mutant into an altruistic population is a case of the same type. These are evolutionary analogs of a prisoner's dilemma. All would be better co-operating, but given that some defect, you must too. So, as Sober notes, being fitter than one's rivals is not to be conceptually identified with being better adapted than one's rivals. You can be fitter in virtue of a change that dooms your lineage to rapid extinction.

Second, as Sewell Wright showed, adaptedness is often explained not by selection alone but by selection and chance. The division of a population into smaller groups so drift and selection can interact is an important mechanism in the explanation of adaptedness, an important explanation of why selection does not hold populations inescapably at local Sober-optima, at which they are less well-adapted.

David Wilson (personal communication) argues for a different take on these examples. We do not have two notions of adaptedness. Adaptedness is just relative fitness. But relative fitness must be defined relative to an appropriate level of the biological hierarchy. A change can increase the relative fitness of some organisms within a group whilst decreasing the relative fitness of that group with respect to others. First, I think Wilson's reformulation obscures an interesting empirical question: what forms of good design can we expect natural selection alone to produce; which require the interaction of selection and drift. Second, Gould's and Lewontin's example arises even in an environment in which there is no metapopulation of groups. In those circumstances, selection acting alone will drive through a tragedy of the commons, reducing the adaptedness of all the organisms in the population, possibly driving it to extinction. In this case, we could not describe the process as one in which the expected fitness of the group changes in the opposite direction of that of the individuals with the new trait. For a group only has a relative fitness if there are other groups.

The heuristic notion of good design is central to the explicitly adaptationist programs of Maynard Smith, Dawkins, Williams, and the Darwinian Psychologists. I think that is fortunate, for an adaptationism that defined itself in terms of claims about the over-riding importance of selection in the evolutionary process would be in serious trouble. Such adaptationism – adaptationism as Sober defines it – involves an implausible claim about actual-sequence explanations. Portraying the panselectionist as believing some analog of O for most interesting traits portrays them as intending to give an actual-sequence explanation of the evolution of a trait in a population, whilst believing that an actual-sequence explanation can safely disregard all factors but selection. I think this is Sober's picture of the adaptationist program; hence he thinks that adaptationist claims are tested by requiring close quantitative agreement between the selective model and observation; qualitative agreement does not suffice (Orzack and Sober 1994). As will be seen, I think adaptationist hypotheses are robust-process hypotheses about evolutionary history. They are hypotheses that claim that some aspects of an evolutionary history are insensitive to disturbance. On that view, our best tests will be qualitative comparisons between related lineages in distinct selective regimes, and comparisons of unrelated lineages in similar regimes.

Once we distinguish actual-sequence from robust process explanation, I think it's widely agreed that if you want to understand the *specifics* of a trait's evolution you must see selection as working together with drift and with historical and developmental constraints characterising the population. If we have in mind the actual sequence of an evolutionary history, it is a mistake to oppose selection to history, development, or drift. For example, in his entertaining paper on kiwi eggs, Gould presents as rivals two explanations of kiwi egg size (Gould 1991). One is the idea that the large egg is an adaptation to ensure the chick is very well developed with good food reserves on emergence; the other tells a historical story about starting point. Kiwis are shrunken moas, and have inherited a moa-size egg (sadly for this story, recent sequence data casts doubt on close moa/kiwi affinities). But here history and selection are not rivals. If we accept Gould's version of the facts of the case, selection and history march in step. Kiwi eggs are large because kiwis retain moa size eggs; selection favoured reduction in overall body size without favouring a reduction in egg size. Had there been selection for chicken size eggs, kiwis would have laid smaller eggs. In this case, adaptation and history merely highlight two aspects of the same process. These are no more rival explanations than, in the causal explanation of a collapse of a star to a white dwarf, we get rival explanations through citing initial conditions rather than the physical laws.

What of other cases? In "panda's thumb" cases, history prevents the evolution of an efficient thumb, so then one might reasonably suppose the effects of history and selection are rival forces acting on the trajectory of the species. Even here, I think this is the wrong picture. For the notion of a selective regime is *itself* a historical notion. On first glance, the echidna and the numbat are living in a common selective regime. They are both ground-dwelling heat tolerant ant-eaters. On this view, their divergent responses to their common environment derives from their different phylogenies. On a more historical conception of their selective regime, they live in distinct environments and under distinct regimes. For their differences modulate their interactions with their biotic and abiotic environment. Any given generation of protonumbats will have inherited a complex of environment, variation, and response and hence selective forces as focused through that mix. That inheritance interacts with selection and chance to produce a new generation from which the cycle begins again.[4] If that is the right picture, even with the panda's thumb we should not see phylogeny and selection as opposed forces. Their phylogeny bequeathed to ancestral pandas both the need for and *the possibility of having* a thumb.

So if you are interested in an explanation that tells the distinctive trajectory of kiwis' evolution, then initial conditions and selective forces collaborated, and it is just a confusion to insist on the overriding importance of one or the other. But equally, there are other good questions that we can ask about this sequence of events; about the robustness of the process, the resilience of the result, and the sensitivity of the trajectory to initial conditions. These are questions about robustness: Under what circumstances would kiwi-like birds have oversize eggs? What is distinctive of those possible worlds? Now we can make sense of the opposition of inheritance and selection. One hypothesis is that large-egged worlds are characterised by shrinkage within some lineage. Their distinctive feature is not a particular selective force. In all of them, some selective forces have been important, but their characteristic feature is a major reduction in size in a lineage. An alternative is to argue that quite different starting points can end in egg gigantism; these evolutionary trajectories are characterised by a distinctive selective regime, whatever it might be. If Paul Griffiths (1994) is right in arguing that adaptive explanations are restricted to lineages, then the adaptationist robust-process explanation is an intersection of these: large egg worlds are worlds in which ratites are exposed to a distinctive selective regime.

These disputes become intelligible seen as debates about the importance and possibility of robust-process explanations of evolutionary process. A typical adaptationist claim within the primate lineage about, for example, testicle size in chimpanzees, is that all large testicle worlds are worlds in which there is sperm competition amongst male chimps and that most such worlds are large testicle worlds. Gould's recent emphasis on the contingency of history makes sense of his scepticism about such explanatory models (Gould 1989, especially Chapter 5). I think we can best understand Gould as arguing that evolutionary history exhibits "sensitive dependence" on initial conditions; very small changes in initial conditions might make very large and non-additive changes in downstream consequences. If very tiny changes in initial conditions or other factors would result in a very different profile of testicle size, then many sperm competition worlds would not be large testicle worlds; there is no such modal information to give.

So we can make sense of opposing history to selection with the kiwi, but not in a way that Gould would find appealing. The "contingency" and context sensitivity of history is just as bad news for *robust historical explanations* as it is for *robust adaptive scenarios*. The sensitivity of the evolutionary process to minor variation undercuts the Mighty Bauplan just as much as it does the Omnipotent Selector. If evolutionary trajectories are sensitively dependent on initial conditions, then many shrunken moa worlds are small egg worlds.

A parallel might make more clear this conception of adaptationism. Sober thinks that "for adaptationists, models that focus on selection and ignore the role of nonselective factors provide *sufficient explanations*" (1993, p. 122). So consider an astronomer's model of the trajectory of the planets in the solar system. An analogy of Sober's sufficiency thesis might be:

(PM) A model of the solar system that considers only the gravitational interaction between a planet and the sun is sufficient to explain that planet's trajectory.

Of course, other forces exist, but they are negligible. Similarly, Sober's adaptationist seems to be saying:

(SA) A model of the evolution of a population that considers only selection is sufficient to explain the trajectory of that population through phenotypic space.

Of course, other forces exist, but in the particular environment in which selection acts, they are negligible.

I suspect few have believed SA. All agree that the phylogenetic inheritance of a population is of great causal importance to its having the trajectory that it has. Suppose that we vary phylogeny, population size, genetic variation, sampling noise but not selective regime. Then the adaptationist believes that all those possible trajectories will share an important property. They will not be the same, even to a first approximation. But they will have something important in common. Armed with the distinction between actual-sequence explanations and robust-process explanations, we can see the adaptationist is not making a simplistic claim about the relative importance of selection. Rather, the central adaptationist claim is one about the robustness of evolutionary trajectories, not the unimportance of phylogeny, drift or development.

3 COMPLEX DEPENDENCE AND CLADE SELECTION

In this section, I argue that there are macro-evolutionary causes. In particular, I want to defend the idea of clade selection. Williams has recently defended clade selection, but he takes clades to be replicators (Williams 1992). Eldredge (1989) has similar views. I shall argue that clades are adapted; they are interactors.

The geometry of clades varies greatly in bushiness and depth. For example, the arthropod lineage exemplified by horseshoe crabs is deep but now very thin; that exemplified by the katipo spider is old and bushy; the beetle lineage is bushier still. How might we explain these patterns? I see three general explanatory strategies in the evolutionary biologist's repertoire. First, aggregation explanations: the pattern is

a mere summation of individual histories. Where an aggregation explanation is appropriate, the macro-evolutionary pattern is a pseudo-
pattern. Second, macro-evolutionary patterns can have individual
level causes. There is a story to be told about why (say) some particular lineage is deep and wide. But the story turns on some special
characteristic of the organism. So this is modest macro-evolution: a
macro-evolutionary phenomenon, but not a macro-evolutionary explanation. Finally, the macro-evolutionary pattern might have macro-
evolutionary causes. This is a bold macro-evolutionary theory. Let me
expand a little on these themes.

(i) Aggregation: there is no explanation of the geometry over and
above a myriad of differing explanations of all the organism-level interactions which aggregate into the contrast in extinction patterns. It may
be, of course, that the pattern we are confronted with is noise; it is the
random variation in extinction of species each of which has an equal
probability of going extinct. But it's important to see that there may
be no macro-evolutionary phenomenon even where species have
unequal probabilities of going extinct. For example, an aggregation
explanation of the extinction of the trilobite lineage would simply
explain the life history of each trilobite species. The extinction of the
lineage is explained by the separate explanations of the extinctions of
all its particular species. There is nothing to be said as a whole about
the fate of this lineage. That might be so, even if some trilobite species
were more vulnerable to extinction (perhaps in virtue of their location
in ephemeral areas) than others. Even so, such a pseudo-pattern does
not support counter-factuals. Suppose someone thought that there was
no common element in the extinction of the eleven moa species. The
fact that all moas are extinct is explained aggregatively, by explaining
the extinction of each moa species. If that were right, had there been
an extra moa species, it would be as likely to be extant as extinct.

(ii) Individual differences can have macro-evolutionary consequences. A particular individual trait may arise through natural selection and result in very different tree geometry. Bushy lineage A might
differ from skinny lineage B in that A's founder evolved parental care.
That behaviour characterises the A lineage and explains its radiation
to a variety of new habitats. The lineage of flowering plants must be a
candidate for an explanation of this kind. So some bushy lineages are
distinguished by "key innovations" which explain their bushiness. (See
Allmon 1992, p. 232, for a survey of alleged key innovations.) In sister
groups of insects, phytophagous clades are consistently more diverse

than their non-phytophagous sisters. A greater population size leads to a better chance of surviving undirected fluctuations. So that if the plant-eating clade began with individual level selection for plant eating, that too would be an individual organism's adaptation that has consequences for the geometry of the clade.

Individual characters that are not adaptations can also explain differences between clades. Extant lineage A might differ from extinct lineage B in an individual trait that is not an adaptation. The trait might be part of the basic developmental system, or perhaps one in which there was no variation, but which explains the difference in geometry of the trees. Consider the contrasting fates of ammonites and pearly nautilids in the end-Cretaceous mass extinction. Embryonic ammonites hatched at only one-tenth the size of nautilid embryos, and formed part of the badly hit surface water plankton. Living nautiluses hatch larger, more developed and in deep water (McNamara 1987). Here we have an individual difference which may well not be an adaptation but which explains the geometry of the lineage. Ammonite extinction/ nautilus survival worlds are tracked by the contrast in their individual reproductive behaviours. Macro-evolutionary patterns with causes like these are all instances of Vrba's "effect hypothesis," of which more shortly.

(iii) Raup, in considering extinction scenarios, considers these ideas (e.g., 1991, pp. 188–93). But there is another possibility he does not consider: properties of the lineages themselves. There are causally important high level properties. The properties of the individuals within a population (their mass, for instance) are causally relevant to population size. But there is no single individual trait that predicts large population size, so it is irreducible to individual behaviour. Moreover, population size is certainly causally important: it may help explain the bushiness of a lineage. For example, tropical organisms are more diverse. A clade that originated in the tropics is more likely to be bushier than its temperate sister group just because of the difference in origin alone. The tropics have greater total energy resources and hence isolated populations are likely to have greater absolute numbers. They are therefore more likely to persist to differentiation. Moreover, size is very important in predicting the outcome of filtering events. But species with large populations do not typically descend from or split into similarly numerous species. So though population size is a causally important property of a collective it is no adaptation. For it is not a characteristic fine tuned by repeated selective episodes. Valentine suggests that in periods of mass extinction high turnover clades are favoured whatever the individual basis of their species' high rates of speciation and extinction (Valentine 1990, p. 139). But unless the species' bases of these turnover rates are inherited through clades deeply enough to face mass extinction events again, there is here no cumulative selection.

Some characters of a clade itself are exposed to many extinction events. Geographical distribution is important both to spread, radiation, or survivorship of clades. This property is obviously important, but is it shaped and maintained in the lineage by repeated episodes analogous to repeated selection, generation by generation, on organisms? If distribution is not honed by cumulative selection, however important it is to the shape of the trees it is, like population size, a causally important property but no adaptation. This is an important question, but I think distribution through both physical and ecological space may well be the result of cumulative selection. Suppose a lineage survived an extinction event (meteor strike, climate change) because it happened to be a little more widespread, through range and niche, than its rivals. It then, as a result of the extinction event, has the chance to move into vacated range and niche, and hence is somewhat more widespread yet. The next extinction event it therefore survives, and the process repeats. Now is this not honing by cumulative selection? And note that we can tell this story about both niche distribution and geographic distribution. These factors, in combination with extinction events, create the conditions for their own increase. The worlds in which lineage A survives extinction events and lineage B does not are those worlds in which A is widely spaced through ecological role and space and B is not.

One of the effects of sexual reproduction is an increase in the variability of a population and hence quicker responses to changes in selective regime. Asexual vertebrate lineages are surprisingly rare and short. Sexual lineages are more "evolvable." Evolvability – the capacity of a lineage to respond to change – will count as a lineage property if, but only if, it is multiply realised. It is an adaptation only if it evolves. If lineages now present are a biased set of the lineages there have been, and lineages with this property are over-represented in the survivors, then the case for lineage adaptation is strong. Schull (1990) asked whether species were intelligent. I used to think he gave the wrong answer; I now think he asked the wrong question. He discusses the way species differ in their capacities for adaptive response to environmental change. So, for example, feral animals re-evolve wild traits very quickly, for the wild traits are genetically masked, rather than eliminated, by domestic breeding. In these circumstances, a species shows the benefit of previous experience; the feral population evolves faster because of its previous history. In Schull's view, it is unmasking memories, not creating new adaptations. He defends a similar understanding of Wright's "shifting balance." Organisms come to have rational expectations about what will happen in novel circumstances though trial and error in a simulated rather than real environment. Thus they avoid the costs of error in their real environment by paying the costs of constructing a representation of that environment. Similarly, the division of a species into many demes allows a species to use trial and error without paying the full cost of error. I think the right way to evaluate these views is to reformulate them as a defence of differences in evolvability between lineages which do not turn on sex. So we can convert his paper into a case for the complex dependence of the evolvability of a lineage on the characteristics of the organisms which compose them. Perhaps the ability of plants to hybridise is another non-sexual realisation of evolvability.[5]

Schull's very abstract characterisation of the gene pool properties in question converges with recent ideas of Dawkins (1989) and Wills (1991). They focus on identifying more specific characteristics of the gene pool which facilitate evolvability. Gene pools are properties of clades; usually, of a species; sometimes, of a closely related group of species. So characteristics of gene pools are clade level properties; they are complexly dependent on the genotypes of individual organisms. It is reasonable to suppose that some gene pool properties are heritable: a descendant species gene pool will have the same property or a variant of it. If Schull, Dawkins, and Wills are right, these are characteristics with important consequences for evolution; they make some evolutionary changes more likely, and others less likely. These characters will be clade level adaptations if they are the result of cumulative selection; if lineages vary with respect to the characters of their genetic systems in ways that affect the bushiness of the lineage, and which therefore fine tune the genetic system. That does seem a consequence of the views of Dawkins and Wills.

For example, both argue for the significance of reorganisations of the genome. In Wills' terminology, there are occasional *potential altering* changes as distinct from *potential realising* changes, the "normal science" of genetic change. Both emphasise that potential altering changes, and their establishment in a lineage gene pool, while consistent with ordinary micro-evolutionary process, cannot be explained by such process. Such revolutionary changes are likely to be of little short-run benefit and partly for that reason the establishment of such a change will be rare. I think they have slightly different conceptions of the nature of the change. Dawkins takes them to be developmental. The invention of segmentation is his main example, an invention allowing massive radiation. Importantly, once invented, segmentation can be fine tuned by cumulative selection, so that, for example, not every segment is exactly the same. Lineages whose gene pool, distribution, and other developmental resources allowed a more rapid differentiation of segments would, on this hypothesis, have been more species rich and deeper than their more constrained sister taxa. Wills conceives these changes as new arrangements within the genome and transcription system itself; changes which predispose new sorts of potential realising mutations in particular directions, but which do not determine the particular embryology of the changes he has in mind. He thus makes the fascinating suggestion that while convergences – for example, in South American mammal fauna – are indeed the result of natural selection, they are *also* the result of the two mammal lineages inheriting a distinctive therapsid toolbox predisposing for some changes but not others (so the dichotomy between history and selection would be again undercut). Like Dawkins, he is keen to show that these genetic specialisms for evolution in a particular direction are the result of repeated rather than single episodes. For example, in his discussion of lineages that survived the Pleistocene ice age he emphasises the repeated practice the lineages would have received in evolvability as they "tried out" new habitats including high mountain and cold winter habitats. Evolutionary toolboxes, potential altering changes can be fine-tuned in this way by selection between lineages. Thus the lineages themselves are adapted, so we are still here in the realm of vehicle selection.

In an important series of papers, Vrba (e.g., 1984a) introduced a distinction between species selection and species sorting. She aimed to give a deflationary interpretation of large scale patterns in the history of life. Her main example is the contrast in species richness between the impala and its sister antelope clade. The impala clade has few and long-lived species; its sister, many short-lived ones. In her view, this contrast is an instance of effect macro-evolution for it derives from the ecological generalisation of the impala and the specialisations of its sisters. From the perspective developed in this section, her distinction between effect macro-evolution and macro-evolutionary patterns with macro-evolutionary causes is an oversimplification. There are three distinctions to draw.

(i) Effects: macro-evolutionary patterns deriving from individual level events. These in turn divide into pseudo-patterns that are mere chance (Raup's "field of bullets"), patterns that derive from key innovations, and those, like the ammonites, that derive from an organism level property of no great individual adaptive significance.

(ii) Patterns that derive from macro-evolutionary causes but not macro-evolutionary adaptations. Population size and population structure may well be irreducible and causally important, but I doubt that they are heritable enough to be macro-evolutionary adaptations.

(iii) Macro-evolution can be driven by the characters of high level vehicles, by genuine high level adaptations, without the suppression or modification of competition between the lower level vehicles out of which they are built. This section has explored cases of this character. Finally, high level selection can involve superorganisms: vehicles composed out of organisms between whom certain types of competitive interaction have been suppressed. Wilson and Sober (1989) argue that slime molds, eusocial insects, and a very special type of community assembly meet these conditions.

4   COULD CLADES BE UNITS OF SELECTION?

Damuth (1985) is not sceptical about macro-evolution as such, but he argues that no phylogenetically defined collective is a unit of selection. They cannot evolve, since their elements form no unity that is ecologically coherent, that is subject to any given selective forces. Except accidentally, the members of a lineage are not in the one environment. So neither families nor clades are units of evolution. For units of evolution are biological assemblages formed of elements that compete, one with another, and it is among these units that selection operates. So, (i) families of individual organisms do not evolve, for families disperse into a number of local populations; (ii) species do not evolve for species are divided into local populations which can inhabit quite distinct environments and which can be subject to quite distinct selective regimes; and (iii) clades do not evolve, for (to an even greater extent than species) the distribution of the species of a clade through temporal, geographic and ecological space is such that they do not compete, inter- act, nor are subject to uniform selective pressures. Damuth is here thinking of evolutionary changes in the shape of the clade by compe- tition between the species of the clade; his point would apply in spades, he would think, to "clade-clade" competition. So he proposes to replace species selection by avatar selection: an avatar of numbats is a local population of numbats.[6]

I am unpersuaded. First, I think that Damuth simply overlooks the causal salience of clade level properties. He supposes "the causal processes of change . . . [involve] . . . only localised ecological entities and the changes seen in genealogical entities such as clades . . . [are] . . . indirect reflections of these" (1985, p. 1138). But just the properties of phylogenetic units that Damuth points to – their varied distribution through different avatars in different geographic areas and niches – may explain why (say) only one branch of a clade is still extant or why one clade has diversified and another has not. Distribution through ecological space is a causally important property. If it is shaped by repeated episodes of selection, it is an adaptation too.

Furthermore, I am unpersuaded by the idea that selective forces cannot act differentially on lineages. Damuth writes "it is difficult to envision agents, dispersed about the planet, that affect only members of a particular lineage, wherever they happen to be, while having no effect upon similar or identical organisms in the same places that belong to different lineages" (pp. 1133–4). The equation between phys- ical proximity and being in the same selective environment breaks down in a number of ways: (i) selective forces need not be spatially local: for example, organisms, wherever they are, tend to have their species-typical parasites; and (ii) we need to be very cautious in assum- ing that selective forces act in the same way on ecologically similar organisms in the same place. For the impact of selection on organisms, groups, species or clades is phylogenetically mediated. Unrelated organisms that appear to inhabit the "same" environment may respond very differently to "the same" selective force. Even if two organisms are phenotypically similar and operating in similar environments, we cannot assume they will respond in the same way to selection. For there may be important differences in their respective gene pools: one but not the other may have potential mates with very different genes. The moral is that we cannot identify selective forces by spatial location (see also Brandon 1990, pp. 64–9).

Moreover, I have two critical points to make about Damuth's rival conceptualisation of hierarchical selective regimes. I think it is going to be impossible to characterise the populations he has in mind. There is no such thing as *the population* sharing a common environment so that all its members are subject to a common selective regime. For selective forces operate at a wide range of environmental "grains." What should we think of as the avatars of New Zealand's South Island rabbit population? For some selective agents, the populations will be quite small: climate varies quite markedly over the island so different populations will be under different selective regimes with respect to winter temperature. Climate impacts on vegetation, but so do patterns of human use, so a vegetation map will not match a climate map. On the other hand, predators (harrier, falcon, stoat, feral cat) are probably rather uniformly spread through the island, so the predation regime will not vary so much. With respect to the predation regime, the whole South Island might be considered a single avatar of the rabbit population. So I doubt that the avatar is, as Damuth supposes, a well-defined natural unit.

I think the same is true of the community and, still more, the community assembly. It needs to be shown that communities really evolve rather than merely change. Sometimes communities will be as ephemeral as demes. In such cases (unlike that of the generations within an evolving population) their reconstitution does not carry enough of the history of the predecessor for the community to evolve. Avatars within a community may disperse to form colonies. But this is not likely to be a community with the distinctive character of the "founding" community. Unless there is enough information transmission from founding to new community, this is not a quasi-reproduction process. Reproduction requires the recreation of the distinctive developmental resources of the old community. Wilson and Sober (1989) and Wilson (1992) describe cases in which these conditions are met. But it's not obvious that these examples are typical. Furthermore, communities do not have natural boundaries, for with respect to different interactions, we will get different avatar counts.[7]

Finally, let me turn to the notion of competition itself. One reason for scepticism about high level competition is that it is hard to think how groups, rather than their member organisms, could compete. Yet evolution is typically conceived as involving competition: the units between whom selection occurs are in competition one with another, and it is this competition which drives selective evolution (see McIntosh 1992). It is hard to fit classic notions of competition to species or clades. But competition is an ill-defined notion. One idea is that units compete when they use substantially overlapping resources in limited supply.[8] From this we get the idea that competition is the relationship of choice between ecologically similar organisms, and the concept of *competitive exclusion* of one species by another with the same resource use profile. Of course, members of one's own species are normally the most similar of all in their resource use. The objection to macro-evolutionary analogs of natural selection then is that this notion does not apply to species or clades for these higher units do not use resources. It's their members that do that.

In reply, first, it is clear that this argument, if sound, does not apply only to *distinctively genealogical* high level units. It would apply as much to avatars and communities as to clades. Second, it is not clear that it is sound. It is far from obvious how to count resource consumers. Does a termite mound consume resources or do only individual termites? Does an organism consume resources or is that confined to its member cells? Moreover, it is not at all obvious that we should characterise competition through the notion of resource use at all: the "resource use" picture does not apply to the classic Darwinian example of two plants competing in an arid environment via (say) their varying tolerances to temperature change. There is no limited resource at stake here, nor even direct causal interaction between the competitors.

Of course, resource use is not the only way competitive interactions have been characterised. Some have contrasted competition with predator/prey and parasite/host interactions by thinking of competition as a symmetrical harmful interaction. The contrast is with predator/prey and parasite/host interactions. These are asymmetric: one party benefits at the expense of the other. This idea of competition is neither pellucid nor plausible. First, some non-competitive interactions are relations of mutual harm; one example is mutual predation at different lifestyle stages. Kookaburras prey on young goannas which in turn prey on kookaburra nestlings. Another example of symmetrical spite is joint support for a predator population. Imagine prey populations reaching their population peaks at different times of year, and hence between them supporting much larger predator populations than would otherwise exist. Each prey species makes things worse for the other, but neither need be competitors. The same seems true of parasites and predators with the same target species: sheep parasites and sheep predators may make things worse for each other by making the target population smaller than it would otherwise be. But are they competitors? Moreover, in exactly what sense are predator/prey relations asymmetric. Even if predation were to always lower an organism's expected absolute expected fitness (for there would always be some chance that it would be the object of predation), it may well increase its expected relative fitness, in virtue of its superior adaptation to a dangerous world than its species-mates.

So I rather doubt that we can define competition independently and think of it as the cause of selection. Our judgements about competition are driven by our conception of the units of selection. Competition is just the relationship between rival interactors' vehicles; we do not derive our concept of the lineages in rivalry from a prior account of competition. So competitors are identified by considering the results of selection: X competes with Y if they are ecologically similar and the increasing success of X excludes similar success by Y. There is no reason to think that this notion is scale-related.

In this section I have considered two objections to conceiving of clades as interactors. Damuth argued that clades are not natural evolutionary units because no selective agents acts on objects as spatially discontinuous as clades. But I do not think selective agents are identified spatially, an assumption on which this objection depends. A second try is the idea that no higher level units compete, and hence there are no competitors between which selection chooses. But the notion of competition will not bear this much theoretical weight; it derives from our conception of selection and evolution, not vice versa.

5 CONCLUSION

Lineages are interactors in and of themselves. They have properties in virtue of which they grow well, or fail to grow well. These properties are not mysterious. No Gaian holism is being peddled here, for the characteristics in question depend on the characteristics of the individual organisms from which the lineages are composed. The return of the species does not mean the exile of the organism. Yet though there is dependence, there is no simple reduction. A lineage may, for example, respond to environmental change because of the variability of its gene pool, or the range of habitats through which it is dispersed.

149

These are not mysterious properties, but neither are they the properties of individual organisms. I do not imagine anyone has ever doubted the causal import of some of these characteristics, but it has been less obvious that there have been high level selective processes which fine tune them and hence explain their existence in a lineage. They are adaptations.

Let me end on an even more speculative note. Seeing the gene as the unit of selection enables us to see new problems, and old problems more clearly. Thinking of clades as interactors may bring with it a similar advantage. For example, in thinking about high level selection, I and others have been thinking about high level adaptation. But selection on organisms does not always increase adaptedness. Relative fitness ought not be equated with levels of adaptation. We should expect to find some macro-evolutionary analog of a "spiteful adaptation," a characteristic that depresses the absolute fitness of a group, species, or lineage, but which depresses others' prospects even more. Could there be lineages with a high speciation rate and an environmental effect which increase *everybody*'s extinction rates? What of other non-adaptive fitness differences: for example those deriving from sexual selection, and selection reinforcing isolation. If there really were high level selective processes, there seems no reason to suppose they would tend to produce only adaptations, and not other fitness enhancers. But what is the macro-evolutionary analog of these? A macro-evolutionary perspective may give us new phenomena to look for, not just new ideas about phenomena we already know of.[9]

NOTES

1. They use the terms "comparative" and "contrastive," but this terminology invites confusion, for "contrastive" has been pre-empted by Dretske 1972 for a use more like their "comparative."
2. Godfrey-Smith (personal communication) has argued that necessity is much too weak a condition, for many things are necessary for the evolution of such balls. They would not evolve in an environment infested by sharp-beaked dive-bombing efficient ballivorous birds, yet the absence of such birds is no function of chimp testicles. Indeed not: if X is necessary for the evolution of T, it certainly does not follow that some function of T is X-directed. But if T's effect of Xing is necessary for the evolution of T, then I think it is plausible to hold that Xing is a function of T. The absence of ballivores is no function of chimp testicles because it is no effect of chimp testicles.
3. Or in the language of Dawkins 1986, "predictable complexity."
4. Robert Brandon, I think, makes a similar point through his distinction

between the ecological and the selective environment. The selective environment is the one that is relevant to explanation through natural selection, and it is constituted by relations between organism and environment.

5. Though introgression (as distinct from polyploidy) is, I am told, quite rare.
6. The evolutionary notion he has in mind is that of the interactor, the adapted unit, rather than the replicator, though this is not the terminology he adopts. For it's interactors that act, and are acted on, in ecological interactions.
7. Damuth does not unequivocally endorse communities as units of evolution. At one point, he remarks that though communities evolve, they are not themselves units of selection; their evolution is the effect of selection on avatars or organisms (1985, p. 1140). However, he then proposes replacing clade selection with selection acting on assemblages of communities. ("community groups," in his terms). That would surely imply that communities were coevolved units and that distinct communities competed.
8. Hence competition is a relationship between interactors not replicators, for resource use is the prerogative of the interactor.
9. Thanks to John Bigelow, Peter Godfrey-Smith, Todd Grantham, David Hull, Philip Kitcher, and Philip Pettit for their comments on earlier versions of this paper. Thanks to Peter Godfrey-Smith for a series of insightful conversations on adaptation and relative fitness, and to Elliot Sober and David Wilson for penetrating reviews on the semi-final version.

7

Darwin's Tangled Bank

I THE EVOLUTIONARY PLAY IN THE ECOLOGICAL THEATRE REVISITED

Darwin's *Origin* ends with a famous image capturing his view of the relationship between ecological processes and evolutionary change. Darwin invites us to consider

> an entangled bank, clothed with many plants of many kinds, with birds singing on the bushes, with various insects flitting about, and with worms crawling in the damp earth, and to reflect that these elaborately constructed forms, so different from each other, and dependent on each other in so complex a manner, have all been produced by laws acting around us. These laws, taken in the largest sense, being Growth with Reproduction; Inheritance...; Variability...; a Ratio of increase so high as to lead to a Struggle for Life, and as a consequence to Natural Selection, entailing a Divergence of Character and the Extinction of less-improved forms. (Darwin 1859/1964, pp. 489–90)

The distribution and abundance of the plants and animals within this community may seem random to us. But Darwin took that to be a reflection of our ignorance and of the complexity of interactions within the community. But though complex, those interactions do determine survival and failure (Darwin 1859/1964, pp. 74–5). Darwin wrote before the appearance of ecology as a recognisable and separate discipline within biology. But his image captures an enormously influential hypothesis about the relation of ecology to evolution. The task of ecology is to characterise the relationships between the organisms of the tangled bank, and between those organisms and their physical environment. This agenda, if carried through, would explain the structure of ecological communities, and in doing so, it would explain the distribution and abundance of organisms. But it would also explain the evolutionary trajectory of organisms by characterising the selective forces impinging upon them.

It would be anachronistic to attribute to Darwin the concept of a niche. But his image leads quite naturally to the classic Eltonian idea that communities are organised by the causal roles of the organisms within them. A particular tangled bank might contain various suites of grasses, shrubs and trees, together with guilds of herbivores, detritus feeders, carnivores, and the like. The adjacent woodlands will contain a somewhat different array of organisms: fewer grasses and shrubs; more trees. Different community types exemplify different causal roles occupied by different guilds of organism: woodlands contrast both with grasslands and with closed canopy forests. Ways of making a living are available in one but not others. These differences have consequences both for the types of organism that can find a place in those communities and for their abundance. The structure of a community constrains the type of organism that can find a place within it. It constrains community membership. Hence ecological interactions are often explored through "community assembly rules"; constraints on the membership of communities. But the structure of a community also limits the abundance within it of any given kind of organism. Thus grasslands cropped by large herbivores provide plenty of opportunity for the way of life exemplified by dung beetles. And it helps regulate the abundance of those organisms: Dung beetles are regulated by the supply of dung.

In recent reflection on evolutionary biology, there has been much discussion of the fact that developmental biology was not part of the Modern Synthesis. There was no-one writing a *Development and the Origin of Species* to shelve with the works by Dobzhansky, Simpson, Mayr, and Stebbings. Equally, ecology was not explicitly integrated within the synthesis through a canonical text. Nonetheless, in high tide of the Modern Synthesis, the relationship between ecology and evolution seemed clear. That view was aptly captured in G. E. Hutchinson's metaphor of the evolutionary play in the ecological theatre (Hutchinson 1965). A given population of organisms experiences a particular environment. The network of relationships between creatures, and between creatures and their physical surrounds, determines (together with luck, good and bad) the success or failure of individuals within that population. It generates the selective forces which drive evolutionary change. Ecology describes the "adaptive landscape" in which different populations evolved.[1]

If we think the evolution of a species is determined by its position in such an adaptive landscape, then Hutchinson's picture of an ecological theatre is apt. For ecology describes the context within which evolutionary change takes place, and isolates the features of that context which determine the evolutionary trajectory of a population. It describes a theatre which is both context and cause of change. In the most optimistic formulations of this idea, an ecologist brings to particular evolutionary plays an array of functional concepts that abstracted away from a play's particular cast. So Diamond, for example, hoped to find "community assembly rules" for avian communities; rules that would predict and explain the presence or absence of birds from particular island communities (see Diamond 1975; for the general program, see Schoener 1986). These functional concepts include general descriptions of an organism's role in an environment: carnivore, insectivore, herbivore; browser and grazer; parasite and parasitoid. They involve concepts about the overall structure and dynamics of environments: the idea of food chains and food webs, and of energy and nutrient cycling through particular ecosystems. They involve the concept of a niche: a particular way of making a living in a community.

This program in ecology fits very naturally with a particular version of evolutionary biology. It integrates very smoothly with a conception of evolution that is not just adaptationist in taking selection to be the main factor explaining evolutionary change, but adaptationist in a particular way. It treats adaptation as a population's *accommodation* to its environment. On this view, the change in a population over time is an evolutionary response to the demands the environment makes on that population. Features of the environment explain features of the organism, via selection. Organisms are moulded to their environments. The classical conception of a niche encapsulates this picture. For such niches exist independently of their occupants; they constrain their occupants, and their occupants can vary in the degree to which they fit their niche.

Hence one program in classical ecology[2] fitted the classical adaptationist program, by providing a functional analysis of biological environments which explained their adaptive dynamics. Elliot Sober distinguishes between two different species of laws of importance to evolutionary theory. *Consequence laws* specify the evolutionary conse-

154

quences of fitness differences in a population. *Source laws* explain the bases of those differences; they explain why, for example, a finch with a thin pointy beak is fitter than one with a more robust, blunter beak (Sober 1984). If this programme were carried through, it would enable us to formulate "source laws" for evolutionary change. An account of community structure that emphasised the functional similarities between communities integrates seamlessly with an evolutionary biology that emphasises the adaptive response of populations to unchanging features of the environment.

This programme had both an ecological and an evolutionary agenda. On ecological timescales, the rules of community assembly explain ecological convergence: the ecological similarity of, for example, grassland communities despite their different phylogenetic components. Grassland communities are similar because they make available an array of similar niches. Thus in all grasslands, grass seeds are an abundant resource, and there will be organisms specialised to harvest this resource. These rules of community assembly impose constraints on species co-existence, and thus determine both successful and failed invasion. Competitive exclusion will exclude some organisms, whereas others will find vacant or poorly defended niches, which they might then occupy. Stoats penetrated New Zealand communities very easily, for they had no competitors for their niche: that of the nocturnal terrestrial predator. Now that they are established, an animal with a similar profile would find it harder to establish. These rules explain the division of resources in a community. Two taxa with overlapping fundamental niches – conditions under which they would survive without competition – may become restricted to those differing areas to which they are best adapted.

Moreover, just as adaptive thinking is often best seen as a research tool that is most important when we fail to find an adaptation we expect, so too with competitive exclusion. We learn most from *unexpected co-existence*. For that can reveal unsuspected ecologically important differentiation. This is often how Hutchinson uses the notion of the niche. So, for example, in his 1965 he gives some apt examples of differentiation between allied species, including MacArthur's famous example of subtle differences in foraging sites amongst his warbler species (Hutchinson 1965, pp. 33–43). Moreover, Weber points out that Hutchinson's famous "paradox of the plankton" and its solution exploit the principle of competitive exclusion as a heuristic tool. The paradox arises because an application of that principle to apparently uniform freshwater environments seems to predict that we should find plankton communities of only a few species. We do not, and the solution is to argue for subtle heterogeneities, and hence niche differentiation, in that environment (Weber 1999).

On evolutionary timescales, these very same processes explain parallel evolution and evolutionary convergence. Australia's bird fauna includes flycatchers, robins, warblers, treecreepers, and the like. These birds are not closely related to those of the Northern hemisphere with the same names, but they are ecologically and morphologically similar to them. The convergence, the idea goes, is a consequence of adaptation to similar niches. This programme explains competitive exclusion's evolutionary trace, character displacement. For related taxa split niche space between them, and thus come to form guilds of related and similar, but sufficiently differentiated, tribes. It explains adaptive radiation, and the re-establishment of broadly similar communities from new components after regional or global mass extinction.

I think it's widely accepted that this conception of the relationship between ecology and evolution is in deep trouble. Within evolutionary biology, Lewontin in particular has been an effective critic of the view that adaptation is adaptation to the environment. In his view, organism and environment stand in a reciprocal relationship. They make one another; these ideas are important to the argument of Section III, and I will discuss them further there.[3] Within ecology, the classic view has been challenged in various ways. Communities might be less stable; less invariant across similar systems; or less independent of the particular species from which they are built, than the most ambitious versions of this programme assume. I discuss these lines of thought from II–IV. In my view, while one version of the idea that community ecology characterises adaptive landscapes succumbs to those problems, the central idea that ecological theory is connected to evolutionary theory through the characterisation of community dynamic survives.

I then shift to a problem that I regard as more intractable, that of scale. Temporal scale has always been seen as a potential problem for the view that community interactions drive evolutionary change. For in comparison to ecological timescales, evolutionary changes are slow. So perhaps the much remarked-upon pattern of stasis, both in the evolutionary trajectory of individual species, and the "co-ordinated stasis" of ecological groups (Brett, Ivany et al. 1996), is a consequence of the fact that selective pressures, while they exist, and while they are effective, are not sustained. Membership and abundance within a community (together with abiotic conditions) determine the adaptive landscapes in which populations find themselves, but the patterns of selective force they generate are too fleeting to effect permanent and substantial evolutionary change. Apparent stasis is actually a wobble around a phenotypic mean. At a high level of resolution, species phenotypes do change, but usually those changes do not sum to a profound shift in morphology or to species turnover. There have been a number of suggestions along these lines, including from Gould himself (Gould 1995). This issue of temporal scale is clearly important, but I shall focus as well on spatial scale.

It is hard to characterise the grain of natural ecological units. The notion of a community is imprecise, and may not be robust or objective at all. But on any plausible way of counting species and communities, there is one to many mapping of species onto communities.[4] Yet the species is the unit of evolutionary change. It follows that there is no simple transition from events within communities to evolutionary changes in species. Instead, I conjecture that evolutionary change fractionates into two kinds.

1. *Red Queen changes* are the result of interactions between species samples in a community: competition, predation, mutualism, parasitism, and the like. Since the character of these interactions and their importance are usually specific to a community, I conjecture that their effects show a punctuational pattern. For only in atypical situations can events within a specific community cause permanent evolutionary change.
2. *Coarse-grained changes* are (mostly) the result of changes to the general physical features of a species' habitat. These factors affect all the populations into which a species is divided in much the same way, despite the differences between communities. Such evolutionary change will often be phyletic; the phenotype of a species will change gradually over time.

In the following three sections of this chapter I pursue the idea that ecological events are translated into evolutionary trajectories via their effects on local community membership and abundance. In the next section, I argue that the problem of contingency, that is, the problem of the unpredictability of local community dynamics, is serious but not so serious as to undermine the view that ecological processes have evolutionary outcomes via their effects on local communities. In Section III, I argue these processes would need to be embedded within a biogeographic and phylogenetic framework, but that is possible and desirable. For example, niches need to be relativised to clades. In Section IV, I take into account equilibrium, and departures from equilibrium, and suggest that the importance of disturbances and other departures from equilibrium does not undercut the basic picture in which local communities are the transmission belt between ecology and evolution. But in Sections V, VI, and VII, we consider spatial scale and the distribution of species and breeding populations through many communities, and then that picture is transformed in basic ways.

II PATTERN AND CONTINGENCY

As I noted at the end of Section I, the classic conception of ecology has faded, both as a characterisation of its own disciplinary agenda and of ecology's relation to evolutionary biology. One potentially damaging response to the classic conception is the idea that ecological processes are not systematic and repeatable in ways that this picture supposes. Darwin took it for granted that there are determinate rules governing co-existence and exclusion on the tangled bank. These days we are not so optimistic. For example, in Thornton's final synthesising chapter on the reassembly of the Krakatau ecosystem, he contrasts contingency with constraint, and his question has become one of degree (Thornton 1996). To what extent do the organisms already established, together with the physical parameters of the environment constrain new arrivals? To what extent is community structure dependent on an arrival and survival lottery, or on fine-grained details of the habitat or the community already established? If community assembly really is a lottery, or if membership and abundance were determined by the fine-grained details of the biological and non-biological environment, then community ecology will not deliver evolutionary source laws by characterising a population's adaptive landscape. For as I argue in *Explanatory Pluralism in Evolutionary Biology*, adaptationist hypotheses are committed to the idea that the evolutionary trajectories they describe are *insensitive* to small variations in background conditions.

The status of generalisations in ecology is deeply problematic.[5] But the threat of contingency is much overstated. The worry posed by extreme versions of the contingency hypothesis is that there are no patterns at all. The thought here is that membership and abundance within a community is sensitive to so many causal factors that we cannot project from one community to another. If community dynamics are sensitively dependent on their current state, then this would be bad news for the idea that ecology, in characterising community dynamics, characterises the source laws of evolutionary change. For adaptive explanations are robust-process explanations. If an adaptive explanation is sound, the outcome it identifies is independent of small alterations of initial conditions and evolutionary trajectories. But if community assembly is contingent, then the distribution and abundance of species within local populations is not robust.

If community assembly is a lottery, or if population sizes within a community fluctuate wildly in response to small changes in circumstance, the classic conception of the relationship between ecology and evolution is certainly in trouble. But is community assembly a lottery? The lottery hypothesis could not contrast more with one important conjecture in ecological theory: there are ecological patterns independent of the particular taxa from which ecosystems are assembled. Plant succession, despite the existence of many exceptions to the general pattern, illustrates the idea. Cleared or otherwise disturbed lands return to forest in broadly similar ways. First annual weeds appear. Annuals invest in fast growth, and produce many small and hence easily distributed seeds. But they do not compete well, so they are quickly replaced by perennial shrubs which invest more in deep roots and stronger, more resistant, structures. These, in turn, are slowly replaced by trees. Trees grow slowly, investing even more in growth rather than reproduction, and they defend themselves chemically and physically against herbivores and other enemies. They often produce fewer but larger seeds. Even though the particular weed, shrub, and tree species differ greatly from community to community, plant succession looks similar in many different regions.

The programme of abstracting away from the membership of specific communities is found most explicitly in the work of Robert MacArthur. He argues:

> We are looking for general patterns, which we can hope to explain. There are many of them if we confine our attention to birds or butterflies, but no one has ever claimed to find a diversity pattern in which birds plus butterflies made more sense than either one alone. Hence, we use our naturalist's judgement to pick groups large enough for history to have

played a minimal role but small enough so that the patterns remain clear. (MacArthur 1972, p. 176)

One of the flagship theories of this programme is MacArthur's own theory of island biogeography. MacArthur developed a theory of equilibrium species diversity that essentially depended only on the size of the island and its distance from immigration sources. As an island community is gradually re-established, the rate at which new species arrive will decrease. For fewer and fewer potentially colonising species remain in the source areas. Furthermore, the species that disperse efficiently disperse early. So as time goes on, those who have not arrived yet are progressively less likely to arrive. Hence migration in of new species declines. Moreover the extinction rate increases. For the more species there are, the more likely it is that one will go extinct in some given time period from chance alone. But the extinction rate is boosted by more than chance: As the number of species increase, the population size of some species already resident is bound to fall, and that increases the vulnerability of these populations to extinction. Furthermore, as these communities become richer, it becomes more likely that the success of one species depresses that of another. This too increases the prospects for extinction; on Anik Krakatau, the establishment of the peregrine falcon seems to have caused the elimination of the hobby. Eventually a dynamic equilibrium should be reached in which the arrival rate matches the extinction rate. The richness of the island ecosystem will depend most heavily on the island's size, for that will roughly measure the diversity of habitats it makes available, and its proximity to species sources, for that will roughly measure the size of the pool of potential migrants.

In 1883 a series of savage volcanic explosions purged Krakatau of all life and physically reshaped the island itself. So it has become a paradigm of a reassembled ecosystem. When life was re-establishing itself on Krakatau, its particular location, the species present in the local bioregion, their capacity to survive in a grossly disturbed habitat, and their dispersal capacities jointly determined a list of potential colonisers. But chance determined the order of arrival of those potential colonisers and no doubt played a role in determining the survival of those that turned up. The theory of island biogeography is committed to the idea that the basic ecosystem structure that develops – its richness or diversity – is independent of these accidents of arrival and timing. Those accidents may determine which goanna or fig species establishes, but they do not determine the fundamental structure and richness of the ecosystem. MacArthur's programme depends on the viability of robust-process explanations of the assembly of island ecosystems from scratch. It requires that every island that is physically similar and which draws on a similar pool of potential colonists ends up with similar communities. If, on the other hand, ecosystem assembly is highly contingent, those that happen to arrive and establish first filter future arrivals either directly, or by modifying the environment so as to exclude potential colonisers. Notice, then, that a contingency thesis about Krakatau is making an important bet: Organism/organism interactions play a decisive role in establishment and abundance, either directly, or mediated by the way one organism modifies the potential habitat of another.

Thornton argues that the Krakatau community is filtered in various ways, by constraints on dispersal, and by biological and physical constraints on establishment. But species vary greatly in their sensitivity to the filters. Goannas arrive early and persist through many changes: they are tolerant of many physical and biological environments. The establishment of goannas on Krakatau is not sensitive to community filters, so goanna presence is highly predictable and robust. The presence of goannas is not sensitive to the community in place on arrival. Fig wasps are an extreme contrast with goannas. They almost invariably depend on a specific species of fig to complete their life cycle, and since any wasp life cycle requires at least two trees for that cycle's completion (for the female lays eggs in a different tree from the one in which she is born) a fig wasp population cannot establish until its host trees are established. Equally, figs cannot set seed without their wasps to act as pollinators. But these *are* extreme examples:

> Within the Krakatau community there is a continuum of species between the extremes (for example, the fig wasp and the monitor) with the median probably lying towards the monitor's end. (Thornton 1996, p. 244)

If Thornton is right, the example of Krakatau does not support an extreme version of the ecological contingency thesis. For example, Anik Krakatau, a new volcanic island adjacent to the fragment surviving of old Krakatau, emerged in the 1930s. For some time, communities established on that island only to be destroyed by further volcanic activity. The evidence about these false starts is not rich, but to the extent that we have it, the three successive communities that established seem to have been fairly similar.[6]

161

III ECOLOGY AND HISTORY

So ecosystem reassembly is probably not so contingent that it is completely unpatterned. Its structure, though, may still depend on the specific taxa that reach it. If community assembly is highly contingent, and we cannot project from one community to another apparently similar one, then clearly no vision like that of MacArthur can succeed. However, the fact (if it is a fact) that community assembly is not highly contingent does not imply MacArthur's programme *will* succeed. Rejecting contingency is necessary but not sufficient for a defence of MacArthur's programme. For there may be robust patterns about communities that are specific to those composed from particular clades. In Chapter 8, I argue that adaptationist hypotheses are best seen as restricted to particular clades.[7] Within the great apes, lack of digestive specialisation selects for complex manual food processing. Within the macropods, arboreal life selects for the use of the tail as a counterweight, rather than an extra point of attachment.

Lewontin claims that organisms and the environment construct one another both metaphorically and literally. If so, then the assembly and organisation of a community will be sensitive to the taxa that compose it. They construct one another literally: as environments, via the selective filters they impose, cause changes in populations and phenotypes, so to do populations physically alter their surrounds. They do so metaphorically as well: evolutionary history ensures that different aspects of the physical environment matter to different organisms. So it may be that grasslands and woodlands dominated by marsupial grazers are organised quite differently from those dominated by placental herbivores. For marsupials, with their different foot structure, impact physically on soils in very different ways. Thus their physical impact on their environments contrasts with that of other large herbivores. Moreover, they have different metabolic demands and hence "see" and respond to different aspects of that physical environment. Consider, for example, how the ability of female kangaroos to arrest embryological development might affect their range size and their ability to migrate between communities. Their ability to carry an embryo internally and in stasis enables them to escape a constraint on migration to which placentals are subject. Thus grassland community patterns in Australia and New Guinea might be quite different from those in Africa, America, or Asia. Relationships between body size and abundance, or between abundance and range size, might be quite different between Old World and Gondwanan communities.[8] History, in determining the biogeography of clades, would have a critical role in fixing the geographical scope of ecological patterns. If the particular taxonomic composition of a community is critical to its assembly and function, then history must play a central role in explaining the nature of biological communities. For no-one denies that history determines the particular species composition of a community.

Thus, consider an example I discuss (Sterelny and Griffiths 1999). Foxes have never established in Tasmania, and thus many small marsupials survive there. The failure of foxes to establish seems to be a consequence of Tasmanian devils preying on their kits. Devils, on the other hand, have probably been excluded from mainland Australia by invading dingos. So the biological character of Tasmania substantially depends on ice age timing. First, the dingo invasion took place when sea levels were rising. So by the time they reached south-east Australia the Bass Straight had reformed and the sea blocked their route to Tasmania. Second, it depends on a particular quirk of fox/devil interaction. So the biological character of a community depends not just on what niches are occupied but also on the species that occupy them. Perhaps we can predict from general considerations the roles a community makes available. There are no large carnivores on small islands. Even so, there will be much that matters about that community that depends on the occupants of those ecological roles; its character will depend in part on particular taxa. Not every devil-sized carnivore would exclude foxes. Similarly, on Krakatau the exclusion (if it is exclusion) of the hobby by the peregrine may have important consequences. For the size range of their prey overlaps but is not identical. The presence of the falcon is bad news for pigeons (Thornton 1996, pp. 199–202), but it might be good news for some small birds or large insects that would have been favoured prey of the hobby but not the falcon. For example, the plain-throated sunbird seemed to be sliding to local extinction after the hobby established but has recovered since they disappeared (Thornton 1996, p. 198).

Thus ecological patterns might depend not just on general physical and geographic factors but also on the specific lineages that dominate a landscape. But such generalisations can still be robust, and hence link ecology to evolution by identifying the selective forces acting on a lineage. If Tasmanian Devils robustly exclude foxes, then we can characterise and explain the within-community processes that have driven foxes to local extinction. It would turn out that the rules of community assembly could not be described in general functional terms: medium size carnivore, small diurnal seedeater, dungivore, and the like. Instead, they would have to be characterised with respect to clades. We should speak (as ecologists often already do) not of woodlands and forests, but of eucalypt woodlands and conifer forests.

In practice many hypotheses about community assembly have been formulated in taxon specific ways: that is true of Diamond's bird co-occurrence rules (Diamond 1975); those about anolis lizards (Roughgarden and Pacala 1989); and Darwin's finches (Weiner 1994). I suspect that many ecologists have hoped that these will prove to be "phenomenological generalisations," patterns in the data that will be explained by a more abstract and more general theory of limiting ecological similarity (Cooper 1998). If Lewontin is right in thinking that organisms help construct their own niche, then this hope is vain. For then community dynamics will depend on the characteristics of specific taxa. For the dimensions of the environment that matter are not physical parameters of general relevance to many unrelated taxa, but dimensions of the environment whose salience depends on the specific biology of the kind of organism in question. Setting aside other mustelids, it's hard to think of real ecological equivalents of stoats. Small cats hunt in quite different ways – they do not, for example, pursue prey into holes and burrows. Rodents, too, are no threat to many animals in real danger from stoats. But clade specific rules suffice to link community ecology to evolutionary theory. As Griffiths and I argue (Sterelny and Griffiths 1999), clade specific rules need not be fine-grained. If rules for community assembly or community function could be formulated in terms of macropod clades, or those of monitors, figs and fig wasps, and the like, our patterns would have broad application over large chunks of the globe and long stretches of time.

IV THE LIMITS OF RED QUEEN ECOLOGY

In 1970, van Valen formulated explicitly an idea implicit in much evolutionary thinking (van Valen 1973). The Red Queen hypothesis implies that the biological environment is of decisive importance in explaining adaptive evolutionary change. If evolution mostly fine-tuned organisms to their physical environment, adaptive change would wind down, as organisms approached optimal design (within the constraints imposed by biological materials) for their habitat. But organisms are always having to cope not just with their physical world but their biological world: their competitors and enemies. They are engaged in a constant race – the Red Queen's race – to maintain parity.

The classic programme is an ecological equivalent of the Red Queen. The focus on niches, and especially the exploitation of the principle of competitive exclusion in explaining the division of resources within a community, exclusion, and character displacement, all bet on the importance of organism-organism interactions in explaining distribution and abundance. That bet might be misplaced. There has been a famous controversy within the ecological community about the importance of competition in structuring communities.[9] Is the distribution and abundance of organisms largely controlled by "density-dependent" factors; factors that become more severe as the population increases? Thornton, for example, in summarising the key question about Krakatau writes:

> If the presence of a species in a community were totally independent of the presence of any other particular species, its establishment on an island after arrival would depend only on the presence of a suitable physical environment. The composition of the community . . . would not matter. If, at the other extreme, establishment were biotically determined, a colonist would succeed only if a particular combination of species were already present. . . . An island community of this type could only be entered by certain species, entry or exclusion being determined by the composition of the community itself. The jigsaw picture of this community could only be extended if the piece to be fitted (the colonist) exactly coincided in shape with the empty space (ecological niche) to be filled, and the shapes of this space would be precisely controlled by the shapes of the surrounding pieces (the species present) already fitted into this picture. Built up in this strictly deterministic way, the community would be a . . . network of interconnected species with strict, tight constraints on which species could and could not enter. Knowing the composition of such a community, therefore, one should be able to predict successfully that a particular species would be able to enter, and for other species the community would be closed. (1996, pp. 250–51)

So if biological interactions were of special importance, the species already present might exclude goannas entirely. If density-dependent biological interactions are significant, the more goannas there are on Krakatau, the harder it will be for any individual goanna to secure the necessities of life. Of course, competition is not the only candidate for

165

an internal biological regulator of population numbers. Disease, parasites, and some prey-switching by predators with no fixed menu are density dependent factors. So even if the importance of competition has been exaggerated, it would not follow that organism/organism interactions are not of decisive importance. In some communities a predator may allow a suite of prey species to co-exist by suppressing population growth in a species that would otherwise be dominant; they function as "keystone species." But even granted that density-dependent regulation does not reduce to regulation through competition, density-dependence may not do the business.

First, the emphasis on biological interaction, especially density dependent biological interaction, seems to be keyed to equilibrium models of ecology. Most obviously, the principle of competitive exclusion is an equilibrium concept. At equilibrium, two species whose resource needs are too similar cannot co-exist within the one community. Yet there is a robust and continuing tradition of thought in ecology that downplays the role of density dependent factors in explaining abundance, a debate that dates back to Nicholson's early work on density dependent population regulation, and the critical response to that work by Andrewartha and Birch, and others (see Kingsland 1985). So there has been an influential tradition within ecology that argues that the classical conception overstates the importance of biological interactions within a community in regulating membership and abundance. Communities may rarely be at equilibrium.

Consider, for example, the "Intermediate Disturbance Hypothesis" (Connell 1978). That hypothesis conjoins two ideas. First, the diversity of a community is maximised by disturbances of intermediate severity. For such a disturbance both creates local environmental heterogeneity and reduces population numbers, thus dampening down density dependent selection. Second, communities are typically disturbed, for the "return time" of disturbance tends to be shorter than the life span of critical elements within a community. Individual trees within a forest will often experience more than one drought, fire or storm. If communities are typically recovering from disturbance, that fact has important implications for the idea that the rules of community assembly and population regulation also define the adaptive landscapes on which a population evolves. For one thing, the intermediate disturbance hypothesis gives weight to those who emphasise the role of events external to a community, in explaining the dynamics of that community. Disturbance is usually defined broadly. Some disturbances will be produced by the community itself. A plague of elephants – a chance spike in the elephant population – might well disturb an acacia woodland. But disturbances are mostly irruptions into a community: fire, flood, drought, and the like. So disturbance explanations are characteristically "outside-in" explanations. What happens within the community is not of its own making.

Second and more subtly, if both within-community density dependent effects, and density-independent external effects are important, then the idea of a single adaptive landscape explaining a population's evolutionary trajectory is in trouble. In Wilson's and Sober's defence of group selection, they argued that an "averaging fallacy" has pervaded much thought about evolution (Sober and Wilson 1998). Suppose we notice that in the last twenty thousand years, red kangaroos in Australia have been on average getting smaller (as, indeed, seems to have happened). It would be a mistake to assume that this must be a consequence of an across-the-board fitness advantage of smaller kangaroos over larger ones. The many local communities through which kangaroos are distributed may vary. In some, small kangaroos are favoured. In others, the large are favoured. The decrease in average size results from the later mathematically swamping the former. If this were so, it would misrepresent this change to say that small size confers on kangaroos a fitness advantage. Sober and Wilson argue that to understand that process, we need to disaggregate the overall fitness figures.

We should also disaggregate across time. Imagine, for example, that two taxa with similar resource needs co-exist in an environment in virtue of their differential levels of adaptation to disturbance and to near-equilibrium conditions. We could calculate fitness values for each, taking into account both types of selection. But it would be a type of averaging fallacy to think of those figures as giving for the two taxa a measure of goodness of fit to their niche. That would not be a good way to represent the ecology and evolution of those groups in those communities. The selective forces acting on a population often push in different direction; perhaps hiding from predators increases exposure to parasites. In such cases, the optimal phenotype in an environment involves a trade-off between these different demands. But this is not a case of that kind. There is no single fitness spike set by the balance of different forces on which both taxa converge. We can explain their co-existence only if we segment the two modes of selection. If selection in near-equilibrium conditions is important, then competition and other density dependent factors within the community play an important role in explaining the survival and abundance of the two taxa, even if it does not play the only important role. But we should separate it from the role of selection in disturbance conditions.

To summarise the argument to date: The most ambitious idea for connecting ecology to evolution through fully general source laws about community assembly and community function seems most unlikely to succeed. But the collapse of that classical programme does not in itself undercut the idea that ecology, in explaining the organisation of communities, provides evolution's source laws. Source laws might vary importantly from community to community, as a consequence not just of differences in physical habitat but also through the characteristics of clades represented in those communities. Podocarp temperate rain forests may well have distinctive features of their own, setting them apart from other evergreen forests. They provide different resources than forests composed of angiosperms, or those dominated by conifers or southern beech. But there may well still be robust generalisations about communities of that type. Similarly, communities may not be at, and may never reach equilibrium, and yet impose a distinctive selective regime on their members. For a population might be selected for particular disturbance regimes, as are many fire-resistant Australian eucalyptus and acacias.

So this general framework for connecting the two disciplines can accommodate departures from equilibrium; a recognition of abiotic and external irruptions into a community and the importance of history. However, the problem of scale poses problems of a very different kind, and transforms our view of the relationship of ecology to evolution.

V SCALE: THE ECOLOGICAL FRACTURING OF SPECIES

The central problem for connecting ecology to evolution is the problem of scale. The critical problem is that the unit of evolution is the species.[10] Yet species are typically distributed through many different communities, and the local populations in those communities often occupy quite different niches. Species are typically *ecologically fractured*. The common brushtail possum is found in communities as varied as cool temperate New Zealand rainforests, inner Sydney suburban gardens and eucalypt woodlands. Thus their relations with those organism on

168

which they feed; those with which they compete; and those which threaten predation all vary importantly from community to community. The concept of a community is a difficult and contested term in ecology (Jax et al. 1998). But on any view of communities, these animals are spread through many. There is no single set of biotically driven selective pressures acting on the possum population as a whole. Their physical environments vary, too. Possums, of course, are exceptionally tolerant and have an unusually broad geographical distribution. But roughly the same is probably true of species with narrow geographic ranges. For even if the geographic range is very narrow, variation in time will often have a similar effect in exposing a population to varying mixes of selective forces. The local composition of pollinating species can vary markedly from year to year (Thompson 1994, p. 184). Lawton's studies on bracken patch communities argue for somewhat more stability in the one patch over time, with species compositions staying fairly stable, as does the rank order of their abundance. But he does note that these similarities are much less marked over periods of twenty years than they are over two (Lawton 1999, p. 181). So even when we consider species with restricted distributions the biological associations of species members may change in important ways over their own life spans or those of their immediate descendants. Typically, species are geographic and ecological mosaics. Species do not have niches.

Moreover, the idea that species are distributed through many niches does not depend on Elton's conception of a niche. Hutchinson reformulated this concept, representing a niche as the region a species occupies in a many dimensional space. Each dimension in that space represents an environmental variable of significance to the species. For example, the dimensions for a woody plant might represent rainfall, canopy cover, temperature, and the levels of various soil nutrients. However, these dimensions do not represent independent causal factors (James, Johnston et al. 1984). In a water ecosystem, turbidity may be an important and positive factor for a species of small fish if predators are present, but unimportant, or even negative, if they are absent.

It is true that organisms modify and select their environment in many ways. A series of woodlands might experience quite different rainfalls. But an insect species might select its living space within these woodlands so that the different populations experience similar environments. The organism's response to its environment might standardise to some degree their experience of their different communities. The selectively relevant environments (Brandon 1990) may be more similar than we suppose. In some cases this selection effect can be quite dramatic. Some parasites are able to go through their whole life cycle on a single organism. When parasites are able to live and reproduce on a single organism in this way, it is usually the case that each member of the parasite species selects a host organism of the same species (Thompson 1994, pp. 124–5). So this parasite life style selects for specialisation. As these organisms increasingly specialised, their selective environment is increasingly dominated by a single feature. Even if the host organisms are distributed through a variety of different communities, the selective environment of the parasite is quite homogenous, and this may help explain the capacity of such parasites to evolve physiological and morphological specialisations to their host. Even so, many species are not ecologically cohesive entities. To the extent that selection pressures derive from ecological circumstances, these selection pressures will not act on a species as a whole.

The fact that species are mosaics has important consequences for connecting ecology with evolution. The distribution of a species through an ecological mosaic, together with gene flow between the fragments, acts as an evolutionary stabiliser; as a brake, *Mayr's Brake*, on evolutionary change. Recent evidence suggests that populations can respond to selection quite rapidly (Thompson 1999, p. 12). Thompson uses this fact to show that co-evolution often involves the reciprocal responses of species fragments to one another. Many apparently generalist parasite species – species which attack many hosts – have turned out to be metapopulations in which the individual populations specialise on one or a few hosts (Thompson 1994, pp. 128–32). But unless that response is linked to speciation, it is likely to be ephemeral. For the gene combinations underlying the new adaptation will be broken up by migration from the parent population.

If species are typically distributed over a number of different community types, and if particular communities are ephemeral on evolutionary timescales, there is no reason to expect natural selection to be acting the same way in different communities. So selection-driven shifts in one community are apt to be undermined by Mayr's Brake: gene flow from other communities will limit the extent of selection-driven response to purely local conditions. Such response as does take place is always vulnerable to dissolution if the local population becomes fully integrated into a larger group. Moreover, if communities disappear and are replaced by others only broadly similar, or if their composition changes in important ways over time, then even if one community type is dominant at a time, it is unlikely to be sufficiently persistent to produce a gradual and entrenched shift in species phenotype. The result is stasis.[11]

The result, then, is that community ecology looks to be an unpromising transition belt for converting ecological events into evolutionary changes. For though local populations may adapt to local circumstances, those adaptations are swept away as communities fragment, combine or change, through the homogenising effect of gene flow between populations. The variability of communities on quite short temporal scales, and the fact that species are typically ecological mosaics, suggests an important hypothesis about the means by which ecological events drive evolutionary histories. Ecological mechanisms operate at two different scales, and produce two different evolutionary patterns. Coarse-grained ecological patterns underpin phyletic evolution. Fine-grained, population scale events produce the famous pattern of punctuated equilibrium: stasis punctuated by change with speciation. So my conjecture is this:

1. Adaptations to the abiotic environment, and adaptations driven by factors internal to the species itself (most obviously sexual selection[12]) often evolve through "phyletic evolution." For such selective forces are often coarse-grained. They are common to all or most of the communities through which a species is spread, and they act in the same way in all of them.
2. A species' adaptation to its biological environment will often show the classic punctuational pattern. For adaptation to the local biota is subject to Mayr's Brake, and hence becomes entrenched only when Mayr's Brake is released. The same is true of local non-adaptive characters that drift to fixation in small populations.

VI COARSE SELECTION

A pattern is not itself a causal hypothesis. Rapaport's Rule – species found at higher latitudes have greater range sizes – describes a pattern, not its explanation. Both the generality of the rule, and its explanation where valid, are issues of active debate. But if our interest is the transmission of ecological events into evolutionary histories, our concern must be with patterns. That is particularly true if our interest is adaptation: there are no adaptive responses to unpredictable events. A chance fire that burns through a forest patch is drift. The mortality it causes is luck, good and bad. A forest patch subject to repeated, though perhaps irregular, burning is subject to selection for fire resistance. The mortality caused by such a fire imposes a selective filter on a population. So pattern in ecology is selection in evolution. Those patterns may not be there to be found in the interactions within a community. Lawton, for example, argues that community ecology studies events at the wrong spatial scales to isolate the critical drivers of evolutionary processes. Thus he argues "locally, at smaller scales, nature will create endless variety, and generalisations disappear, or at least become more difficult to see" (Lawton 1999, p. 179). Contingency – the sensitivity of ecological outcomes to many causal factors – is a problem for every ecological enterprise. But community ecology, Lawton suggests, is likely to be overwhelmed by this contingency. For *so much matters* in the determination of the composition of, and abundance in, local communities. Hence the scarcity of robust generalisations in community ecology. Moreover, contingency implies variability: The make-up of, and abundance in, communities will vary on short temporal and spatial scales.

Lawton's scepticism about the existence of robust generalisations about the internal dynamics of communities does not extend to ecology in general. To the contrary, his point is that patterns appear at larger geographic and temporal scales; regions and processes that affect many communities. One instance he notes is Whittaker's attempt to predict plant community types. Whittaker isolates a range of plant community types[13] and argues that to a reasonable approximation the formation of those community types and their geographic distribution is controlled just by mean annual rainfall and temperature. Moreover, often, the members of these community types have an array of characteristic adaptations which are responses to these coarse environmental variables rather than the fine structure of the interactions within specific communities. Desert plants, as he notes, often have deep and wide-ranging roots for effective water uptake; they have water storage tissues; their leaves have a characteristic range of adaptations to reduce water loss; in some cases, even the chemistry of photosynthesis is adapted for higher temperatures and less water loss (Whittaker 1975, pp. 167–9).

Perhaps palaeoecology tells the same story. The phenomenon of co-ordinated stasis is sometimes interpreted as the persistence of an ecological assemblage – not just of the same species set but the same species with stable relative abundance and with minimal morphological change – over evolutionarily significant periods of time. On one view, these assemblages attest to the effect of regional, larger scale ecological patterns stabilising variation at the level of individual communities (Morris, Ivany et al. 1995). In my view, though, we should interpret this evidence of stasis with considerable caution. For where our evidence is best – with contemporary and recent ecologies – we see much less evidence of this ecological stability. The "Pleistocene Paradox" is precisely the evidence of fluctuation in community membership evident in the high resolution record of the recent past (Brett, Ivany et al. 1996, pp. 11–12).

Along similar lines, Ricklefs, Lawton, and others argue that community richness is mostly determined by external coarse-grained factors. Instead of richness being determined by the intricate jigsaw puzzle of biotic relations within a community, it is driven by regional species richness – the set of species available to make up the community – and the physical features of the habitat (Ricklefs 1989; Zobel 1997). Thus when Krakatau's ecology was being reassembled, vegetation mostly arrived via the sea, and was deposited at or near the tideline. So obviously only water and salt tolerant plants had any chance of establishing. Nothing that required a biologically rich soil structure was a potential member of those early communities. The suggestion from "macro-ecology" is that robust, and hence relatively invariant, ecological processes occur at coarse physical scales. They occur at the scale of habitats, not local communities.

The patterns detected by macro-ecology are relatively independent of the specific structure of communities, and interactions within them. One obvious source of such pressures is widespread climate change. The browning of Australia – as it slowly became hotter, drier, and less fertile – ought to have generated selection pressures on the right temporal and physical scales; pressures that were fairly independent of the specific details of community structure. The whitening of Antarctica is another example, but its details are probably irretrievably lost. So we might expect to see Gondwanan palaeobiology as a test of punctuated equilibrium. It seems to be an arena in which "phyletic gradualism" – slow evolutionary change within a lineage – might produce adaptation to the abiotic features of the environment. Naturally coarse-grained features need not be abiotic. Cane toads have penetrated numerous Australian communities, and presumably affect them in broadly similar ways. An adaptation to cane-toad presence in one community would be adaptive in many.[14] Nonetheless, as I noted above, most factors that affect most or all of the communities through which a species is spread are features of the physical environment.

In short, there are some biological factors, mostly deriving from features of the physical environment, that operate over larger scales than that of a local community, and which affect organisms in ways that are largely independent of those organisms' interrelations with other organisms. When the scale of a biological factor includes the whole range of a species, and when it exerts a selective force on the members of that species, the fact that the species is an ecological mosaic will make no difference to its response to selection.

VII COMMUNITY EVOLUTION: RELEASING MAYR'S BRAKE

In Section VI, I argue that some ecological processes act on spatial scales much larger than those of a community, however we define this difficult concept, and in ways that are relatively independent of the specific dynamics within a community. Plausible examples are gradual, large-scale climate change. But tectonic and other geological forces that alter the basic structure of the landscape (uplift; erosion and deposition; sea level changes) will probably exert similar coarse but consistent effects over large areas.

Such changes have the potential to have another effect. They can release Mayr's Brake by simplifying the relationship between species and environment. Community interaction *can* generate permanent adaptive change on rare occasions. The ephemeral nature of communities and the flow of genes between them mostly prevent the entrenchment of local adaptation, but they do not always do so. In somewhat different ways, Eldredge, Vrba, and Bennett have investigated the circumstances under which community interactions become motors for permanent, speciating evolutionary change (see Bennett 1997; Eldredge 1995; Vrba 1993; Vrba 1995). For climate change and other large-scale physical changes can turn species mosaics into patchworks of isolated populations. For example, Vrba argues that the distribution of a species is often insensitive to the presence or absence of other species. Hence they can live in many different communities. But all are sensitive to key physical parameters: temperature, moisture, sunlight and the like. She argues that each species has a distinctive habitat, defined by the range of physical parameters in which it can survive and reproduce. These parameters are often sensitive to climate changes. So what happens when the climate does change? Sometimes the effects on the species' range will not be dramatic: The potential space available to a species might shrink a bit, expand a bit or shift latitude. If physical barriers do not intervene, the species can shift with it. Sometimes the potential space will shrink to zero, and the species will disappear with it.

However, stasis breaks down when environments both change (creating new selection pressures) and a species' range fragments, dissolving the metapopulation by chopping it into its component populations. Thus external shocks to the system on regional scales can both simplify the relationship between the evolutionary and the ecological units, and change the pattern of selection that operates. These shocks can create the conditions under which selection can be effective (releasing Mayr's Brake) while causing it to act. A local, isolated population is not ecologically fragmented. The vast majority of such small populations will go extinct. But if Bennett is right, orbital oscillations generate climate change in cycles of twenty thousand to one hundred twenty thousand years. Thus widespread species mosaics will often be cut into isolated, ecologically homogenous fragments, repeatedly creating the potential for evolutionary response linked to speciation (Bennett 1997, pp. 185–95). In these fragments, the population is not so small that genetic variation is sharply reduced. Selection can act, and act without counter-balance by homogenising gene flow from neighbouring populations. For there are no neighbouring populations. Even if most such populations go extinct, or fail to evolve isolating mechanisms, a few will survive as new species. Thus Bennett has formulated a more general version of Vrba's "turnover pulse" hypothesis. A turnover pulse takes place when species from quite different lineages and with different ecological profiles disappear from the fossil record at around the same time. In their place appear new species closely related to the departed. She thinks she has detected pulses of this kind in the African fossil record, and she thinks it is caused by the simultaneous fragmentation of many species' habitats, producing many isolated populations, each likely to go extinct, but each a potential new species. Notice, though, an asymmetry. The turnover pulse could produce adaptations to changed physical conditions. However, if my conjecture about two evolutionary modes is correct, turnover pulse conditions are not needed for abiotic adaptations, but they are needed for adaptations to the local biological conditions.

VIII CONCLUSION

Time to sum up the state of play. At its climax, the relationship between evolutionary theory and that of ecology would have seemed transparent. Ecology was to characterise the organisation of communities in such a way as to expose the selective forces acting on populations. In the last few decades, that picture has been transformed from both sides. Richard Lewontin, in particular has pointed out that organisms do not just accommodate themselves to their world; they select and transform their environments as well. Stephen Jay Gould has argued that the synthesis tacitly accepted an extrapolationist view of evolutionary dynamics, a view treating evolution at the level of species as a simple aggregation of evolutionary changes within populations in communities (Gould 1995). But once we appreciate the fact that species are ecological mosaics, distributed across many communities, it is obvious that there is no simple translation of a change in a population within a community to a change in the species as a whole. Equally, our conception of ecology has changed too: communities are probably less predictable, less often at equilibrium, and more strongly influenced by the specific taxa they contain, than once was supposed.

Some of these theoretical changes in ecology and evolution stretch the classic conception of the relationship but do not fracture it. I have argued that the decisive fact is the mismatch in scale – the fact that evolutionary units do not map onto communities – that really forces a major overhaul. That would pose an intractable problem even were communities at equilibrium, with a transparent functional structure, and hence with predictable dynamics. However, there are some ecological processes that are both predictable and patterned, and whose scale is coarse enough to map onto evolutionary units. Large-scale physical changes – changes in climate, changes in landscape structure – both occur over the whole of a species' habitat, and affect populations within that habitat in a constant enough way to be drivers of evolutionary change. More fine-grained ecological processes – those specific to particular communities – can act as drivers of evolutionary change only in special circumstances. They do so only when migration into a community (and its successors in the same geographical location) is blocked for long enough for reproductive isolation to be established. Hence, we can expect to find the punctuated equilibrium pattern in the biological record alongside evidence of gradual phenotypic change. Gradual change is produced by coarse-scaled, community-independent, mostly physical, factors. The punctuated pattern is produced by community-dependent, mostly biological factors, in those rare circumstances in which a species is fragmented, and a fragment survives to speciation.[15]

NOTES

1. The idea of adaptive landscapes was originally introduced as a representation of evolution in which the dimensions were gene frequencies. But the idea evolved into an ecological representation in which height represents fitness and all the other dimensions represent aspects of the environment (Depew and Weber 1995, pp. 293–7).
2. Only one: this was never the only game in town. See Kingsland (1985) for a fine overview of the permanent tension in ecology between those that emphasized the particular and idiosyncratic nature of ecosystems and those that aimed for general functional analyses of this kind.
3. See (Lewontin 1982, 1983, 1985). These ideas are developed further in Odling-Smee, Laland et al. (1996), Laland, Odling-Smee et al. (forthcoming), and Jones, Lawton et al. (1997). For discussion, see Godfrey-Smith (1996, Chapter 6) and Sterelny and Griffiths (1999, Chapter 11).
4. See also Damuth (1985) and Eldredge (1989).
5. For reviews, see Cooper (1998); Gaston and Blackburn (1999); Lawton (1999); and Weber (1999).
6. Thus Sterelny and Griffiths (1999) now seems to me to somewhat overstate the importance of contingency.
7. The argument is developed in more detail in Sterelny and Griffiths (1999, Chapter 10). See also Cooper (1998). He argues that a causal generalisation in ecology can be "resilient" – a probabilistic version of a robust process explanation – yet quite sharply restricted in scope.
8. Gaston and Blackburn defend macro-ecology but argue for more macroecological investigation into what they call the "anatomy" of general patterns. In other words, they argue that we should disaggregate these patterns into their components, seeing how broad qualitative patterns like size/abundance relations are built from particular cases (Gaston and Blackburn 1999). In their terms then, I am suggesting one aspect of the anatomy of broad ecological generalisations may be phylogenetic.
9. Well-reviewed in Cooper (1993).

10. Perhaps this is true only of metazoan organisms. It is not clear that there is a theoretically robust notion of species that applies equally to the metazoa, to plants, and to unicellular asexually reproducing organisms.
11. In the literature on co-ordinated stasis, there is some suspicion that it may be partially a statistical artefact. It may characterise the evolutionary trajectory of common rather than rare species. Mayr's Brake may particularly apply to widespread and abundant species, for such species are distributed through many communities and migration between them will be frequent. Needless to say, testing the hypothesis that rare species are more evolutionarily labile than common ones is horribly difficult (Ivany 1996).
12. Though the way sexual selection acts can be profoundly influenced by natural selection, and by variations in selective regime across a species' range.
13. These include: tropical rainforest; tropical seasonal forest; savannah; thorn scrub; taiga; temperate shrubland; and so forth.
14. Though this would not solve the problem of a co-adapted local complex being broken up by gene flow, if a temporally isolated population were rejoined to others.
15. Thanks to Roger Buick for his comments on an earlier version of this chapter and for calling my attention to the literature on co-ordinated stasis. Thanks to Geoff Chambers for his comments on a much rougher ancestor of this material.

Part IV
The Descent of Mind

8

Where Does Thinking Come From?
A Commentary on Peter Godfrey-Smith's
Complexity and the Function of Mind
in Nature

I THE ENVIRONMENTAL COMPLEXITY HYPOTHESIS

Plants live in a world of hard choices. Their world abounds in plant-eating fauna of various shapes and sizes. Against these, they have at their disposal an array of defences. But these are not cheap. They divert resources from growth and reproduction. It's a trade-off. Some plants are growers; others defenders; still others try a little of both. An alternative to all these is flexible or plastic response; to grow unless the pests are particularly bad, in which case defend. Common corn is mostly a grower, investing little in defensive chemicals. But if it's damaged by certain bugs, corn generates and secretes a terpenoid mix that attracts the parasitoid enemies of its enemies (Turlings et al. 1990).

Complexity is an investigation into the evolution of adaptive plasticity; in particular, cognitively mediated plasticity. Cognitive capacity is a means to adaptive plasticity, for it enables organisms to vary their behaviour. Godfrey-Smith explores the hypothesis that the function of cognition is to fit organisms to a specific kind of complex environment. There is no point in modulating behaviour unless the environment is heterogeneous in ways that matter to the organism, and to which the organism can respond. If your environment is always infested with herbivorous bugs, there is no point in taking special precautions on some particular occasion. Moreover, environmental variation must be detectable by the organism. There must be some proxy, some symptom that the tree can use as a sign of herbivore abundance. If there is no such sign, the tree's world is intractably complex. The tree must settle for a single best-overall behavioural response. So if the function of cognition is to vary behaviour in response to variation in the world, three conditions must be satisfied.

181

i. The organism's environment must vary in ways that matter to that organism.
ii. The organism must have potential variation in its behavioural repertoire; different behaviours have different payoffs in different environments.
iii. The organism must have access to information about its specific environmental state.

Moreover, though it's necessary that doing the right thing at the right time is better than doing the same thing all the time, that is not sufficient for plasticity to be of value. Information will never be perfect, so trying to do the right thing at the right time risks doing the wrong thing. In high bug times, investing heavily in anti-bug warfare is better than investing modestly. But it's worse in low bug times. So if the cost of bug protection is high, either the signal that you need it must be accurate or the benefit of having it when you need it must be great.

Thus the environmental complexity hypothesis is the idea that cognitive capacity evolves as a selected response to detectable environmental heterogeneity. Godfrey-Smith locates this explanation of cognition in a broader framework, treating it as a special case of two important explanatory families. It is an adaptationist and externalist explanation of cognition. The explanatory arrow is "outside-in." The behavioural complexity of the organism is an adaptation to environmental complexity. One main theme of the book is to explore explanatory patterns of this kind, for both adaptationism and externalism are controversial. Lewontin is well known for his rejection of adaptationist externalism, arguing that organisms construct their environment rather than being shaped by it (Levins and Lewontin 1985). In response, Godfrey-Smith distinguishes sharply between an organism's physical transformation of its environment (attracting parasitoids) and its accommodating itself to its environment (manufacturing deterrent poisons). Large organisms have greater thermal inertia than smaller ones, so they have to worry less about certain temperature fluctuations. Even so, thermal variability is an objective, organism-independent feature of the environment. Features of organisms make some environmental properties relevant and others irrelevant, but they do not make the properties themselves. The idea that organisms construct their environments, and hence cannot be seen as adapted to them, looks plausible only if we lump physical transformation and accommodation together.

Furthermore, we should distinguish *asymmetric* from *interactionist* versions of externalism. Spencer defended an asymmetric explanation of cognition. In his view, the environment develops independently of organic response to its complexity. But externalism does not have to be asymmetric; the complexity of the environment can be in part the causal product of the organism. For example, I read Povinelli (Povinelli and Cant 1995) as defending an interactionist externalist explanation of the evolution of a self-conception. He argues that as arboreal organisms become larger, their world becomes more complex. Small arboreal organisms can follow line of sight routes through their environment, for most branches rigidly support their weight. For them, movement through the treetops is no more problematic than movement on the ground. An increase in size changes all this. Direct line of sight routes are often unsafe for larger animals. Their options are further reduced because they often must support themselves from more than one branch at a time. Yet since branches bend under their weight, routes with apparent gaps are actually possible. So route planning becomes more complex, and their environment becomes more challenging, as a result of a feature of the organism itself. At the same time, as organisms become heavier, falling becomes more deadly, making trial-and-error learning more expensive. Putting all this together, an independently caused increase in size leads to an increase in environmental complexity – in part as a result of organism/world physical interactions – which in turn leads to the evolution of a self-concept. So an explanation of cognition can remain adaptationist and externalist while not seeing organisms as passive clay moulded by autonomous environmental forces.

In discussing the environmental complexity hypothesis, I focus on three issues. First, I argue that it's very difficult to formulate a plausible form of externalism with any real bite. So I suggest decoupling externalism from adaptationism. Second, I explore problems that arise out of testing the environmental complexity hypothesis. I finish by discussing the broader epistemological themes Godfrey-Smith develops from these ideas.

II EXTERNALISM

I am an adaptationist about cognition, and I am sympathetic to its externalist versions. Nonetheless, it's extraordinarily hard to formulate an externalist thesis that is both plausible and which has real bite. So in this section I will air some sceptical worries about externalism, seeking to decouple it from adaptationism. The prima facie problem for externalism is that both external and internal changes are necessary for an evolutionary response to complexity, but neither are sufficient. Hence though externalism is a historically important pattern of explanation, it is hard to see how it could be the right view of evolutionary change. A similar problem is famous from developmental biology. It's often supposed that, for example, the human thumb is genetically caused, even though a host of environmental factors are necessary for its growth. The intuitive idea is that though environmental features are important, they only provide general support to this developmental process. The *specific* developmental outcome is down to the relevant genes. First, the developmental outcome is buffered over a range of environments; it is robust over generous variation in environmental resources. Moreover, when those limits of variation are exceeded, the result is developmental failure, not an alternative structure. So though variation in genetic structure could cause distinct structural outcomes, no variation in the foetus's food, water, oxygen uptake, or protection from thermal variance, will turn the thumb into a useful ornithoptorian claw. So though the thumb's development depends critically on an appropriate environment, these environmental resources are mere "fuel for development." In the relevant genetic structures, the tolerance of variation is less, and the possibility of an alternative structural outcome is greater.

Even in developmental biology, the distinction is very controversial. (See for example Gray 1992; Oyama 1985; and Chapter 3.) But whatever its merits in development, I doubt that it can be recruited to defend an externalist explanation of the evolution of plasticity. Internal factors are structuring causes. They determine whether the environment is relevantly heterogeneous. Internal features of a plant lineage help fix its potential strategies. For instance, they will predispose for different kinds of response to herbivores: direct chemical defence, or mutualisms with ants rather than parasitoid recruitment. Internal features help determine whether the environment is relevantly heterogeneous, and whether there is a right option in response to that heterogeneity. Moreover, particular adaptive response requires particular genetic variability. Genetic variability is not fungible, like food or money, but specific to specific outcomes. So the array of internal factors that are necessary for the evolution of adaptive flexibility are not mere general "fuels for evolution." They are specific, not general causes of evolutionary response. They are causes of structure, not just generalised support for evolutionary change.

Godfrey-Smith is aware of this problem, and suggests two ways that an externalist view might be vindicated (pp. 52–5). The first takes up the adaptationist element of the environmental complexity hypothesis. Sober (Orzack and Sober 1994; Sober 1993) argues that adaptationism would be vindicated inductively if selection-only models of many different evolutionary episodes fit the data well. Godfrey-Smith endorses the idea that an accumulation of specific adaptationist triumphs gradually vindicates a general adaptationist conception of evolution. In Section III, I argue that this is an unpromising characterisation of adaptationism when our focus is a broadly defined category like behavioural flexibility. Moreover, I think it's possible to defend adaptationism without externalism, so vindicating the coherence of adaptationism does not vindicate the coherence of externalism. A second idea is to characterise the externalist adaptationist as making an explanatory bet.

> Adaptationists bet that spending your time and energy investigating patterns of natural selection at the expense of genetic factors is better than spending most time and energy on the genetic system and less on patterns of selection. Adaptationism involves preferring having a model with four environmental factors and two genetic factors, to having a model with the investment switched. (p. 53)

But *how could* an externalist bet be better, if both factors are equally necessary?[1] In a very penetrating discussion of Lewontin (pp. 145–57), Godfrey-Smith defends the objectivity and autonomy of the physically defined environment. The environment exists independently of the organism, but features of the organism determine which features of the environment are relevant. On the face of it, this seems to defend the objectivity of the environment at the expense of the externalist bet. If properties of the organism make some aspects of the environment causally relevant in explaining the organic system's response, we cannot predict evolutionary response just from our knowledge of the environment and a few general features of the organic system. We need to understand the organic system in enough detail to understand which features of the environment will be relevant to it. Without knowing a lot about wedgetail eagles, we cannot predict the impact of the rabbit-friendliness of the Australian environment on their evolution.

Godfrey-Smith's response is to suggest that the features of the organic system which determine which aspects of the environment are relevant to it are not those the externalist explains by appeal to the environment. Here the Povinelli hypothesis works well for Godfrey-Smith. The size of a male orangutan make the load bearing capacities of wood relevant. Size makes its world relevantly complex. But this complexity has not driven the evolution of size. So size explains complexity and in turn complexity explains, externally, cognitive change. However, cognitive evolution is likely to be typified by environment-organism cycles. Think of the evolution of colour vision. Colour vision made colour relevant, and not only began a process in some lineages leading to very fancy colour vision, but also to a changed and varied colour environment. Adaptive explanation depending on these cycles does not seem much like externalism. No explanatory bet is being suggested that focuses most on external factors. No model of construction is envisioned in which features of the environment are specified richly and in detail, whereas features of the organic system are specified with low resolution. There is no suggestion that internal features of the organic system are just generalised support for adaptive change. So it's still hard to see an externalist thesis here worth fighting for.

There is a rather more direct challenge to externalism, one which resists the distinction between internal and external factors altogether. At first glance, scepticism about a principled internal/external boundary looks absurd. Organisms are enclosed by physical structures. These barriers are essential for the metabolic integrity of the organism. From the point of view of physiology, the organism/environment distinction is both sharp and important. It does not follow that the boundary is principled when we consider development and evolution. Oyama is sceptical about the importance of the organism/environment interface in developmental biology because she does not think development is driven by discrete chunks of information, some of it internal and guiding the development of genetic traits, and some of it external, guiding an organism's learning. In her view, the information needed in the development of every trait is constructed from both internal and external sources. Oyama might be wrong, but she cannot be refuted by appeal to the boundary's importance for metabolic integrity.

If the existence of a principled boundary is an open question in developmental biology, it is even more open in evolutionary biology. The metabolic argument for the boundary has even less grip. For the "organic systems" in question do not have skins. For when Godfrey-Smith talks of organisms, he has in mind not single organisms but organism lineages. For example:

> ... the organic system in question does play a role in determining whether ... a given environmental pattern is relevant to it or not. ... But the properties of the organic system that make the environmental pattern relevant need not be the same properties that the environmental pattern can help to explain. ... The organism, by virtue of one set of organic properties, makes it the case that a given environmental pattern is relevant. (p. 155)

Lineages do not have skins. They have no metabolic integrity to preserve. It is not at all obvious that there is a principled demarcation in the causal history of an evolving lineage into external and internal factors. For example, Godfrey-Smith counts game theoretic and other frequency dependent effects in evolution as external factors. But these equally well can be thought of as internal to the lineage; they are after all features of the evolving population. So the set of evolutionary causes may not divide in any clean way into internal and external factors.

I think it's even clearer that adaptive consequences do not divide cleanly into the internal and the external. In my view, this is vividly illustrated by the recent revival of group selection (Wilson and Sober 1994). We can see social evolution as producing adapted groups, or see it as producing individual behaviours that are adaptive only because of the social environment. For example, we have two equivalent explanations of the tolerance of female lionesses of one another's cubs. On one view, the organic system is the pride itself. The greater fitness of suckling-tolerant groups outweighs the within-group fitness advantage of free-riding. Suckling tolerance is an adaptation of the lioness pride. On the other, the organic systems are individual organisms, and the rest is environment. The fitness of individuals in the groups with altruists is higher than the fitness of the individuals in the selfish groups. Two elements combine to explain an organism's fitness. There is the contribution to the fitness the organism has from being in a group of a certain kind. There is a second contribution from its role in that group. Both pictures recognise the importance of the division of the population structure into prides, and recognise that an organism's fitness depends both on the character of the group it inhabits and its character in that group. In my view, these pictures are equivalent (see Chapter 4). If so, there is no objective organic system/environment boundary.

Dawkins (1982) and Hull (1988) also argue against a general, principled organism/environment boundary in evolutionary theory. Dawkins thinks that adaptive gene effects – the effects in virtue of which they are replicated – are often not effects on the body in which they reside. The adaptive effect of the corn terpenoid gene is parasitoid recruitment. Godfrey-Smith treats this idea as a rival view of the organic system/environment boundary, not as a denial of that boundary's significance (pp. 48–50, 137). That is not how Dawkins himself sees his enterprise. He thinks it shows the idea of the boundary is arbitrary. So Dawkins says of the "Bruce Effect," the effect through which a male mouse induces abortion in the female, that:

> Whichever link in the chain a geneticist chooses to regard as the "phenotype" of interest, he knows that the decision was an arbitrary one. He might have chosen an earlier stage, and he might have chosen a later one. So, a student of the genetics of the Bruce Effect could assay male pheromones. . . . Or he could look further back in the chain. . . . Or he could look later in the chain . . . outside the male body. . . . (1982, p. 230)

It would be wrong to read Dawkins as defending a general scepticism about the organism/environment boundary. Dawkins' concept of the vehicle is an attempt to capture what is *distinctive* of the invention of the organism. So for him the vehicle is a *special case* of interaction. It's that minority of cases in which the adaptive phenotypic effects of genes combine together in the design of a single structure in which they reside. So we should see the extended phenotype and trait group selection as decoupling adaptationism from externalism. Godfrey-Smith argues that Lewontin's conception of organic construction of the environment conflates two importantly different evolutionary processes: accommodation to a feature of the environment with transformation of it. I think the same is true of imposing an organism/environment boundary in thinking of adaptive change. It conflates examples where adaptive response does involve the construction of an integrated system behind a physical barrier, with adaptive response which does not have this character.

III ADAPTATIONISM

Thus I think it's possible to be an adaptationist about cognitive evolution without thereby being an externalist, without seeing the causal route as outside-in. But should we be adaptationists, and what is it to be one? As I have noted, Sober takes adaptationism to require models with a tight quantitative fit with observation. This criterion may well be sufficient, but it had better not be necessary. The models Godfrey-Smith discusses do not specify the actual evolutionary trajectory of any real population, but instead attempt to capture an important common element in many trajectories. They are not models that we could expect to deliver close empirical fit to real cases. So I read the adaptationist as defending a different kind of robustness thesis, as defending a qualitative rather than a quantitative claim. Suppose that we vary phylogeny, population size, genetic variation, sampling noise but not selective regime. Then the adaptationist believes that all those possible trajectories will share an important property. They will not be the same, even to a first approximation. But they will have something important in common.

But what is the "something important" and how do we confirm its existence? Given that quantitative fit is beyond our reach, how can the adaptationist go beyond "how possibly" explanations for the evolution of plasticity? It seems to me that we can do this only by synthesising adaptationism with phylogeny.[2] So, for example, if the moderate version of the environmental complexity hypothesis is right, features of the organism lineage make aspects of the environment relevant, and these in turn have further evolutionary consequences. That hypothesis has implications for the pattern of traits in a clade. The traits which make a certain kind of heterogeneity relevant should arise deeper in the tree, and hence should define a broader group within which plasticity nests. On a more local scale, in explaining flexible defence by corn, we would need to chart the distribution of parasitoid attraction in corn's sister group, to try to determine the evolutionary sequence through which an adaptive cluster evolved. We would want to know whether its relatives always signal or never signal; whether army worm predation is a variable feature in non-signalling relatives' environments; and so on. In effect, we would be searching for correlation between the alleged adaptive cause and the adaptive response within a group where the historical and developmental constraints are relatively constant.

Combining phylogeny with a hypothesis about the origins of plasticity raises two issues. First, there are more or less historically nested versions of adaptationist hypotheses. Consider ecological succession. Roughly speaking, as a disturbed habitat becomes colonised, successive plant communities shift from being growers to being defenders. Now suppose you survey succession in one patch, noting the growers, defenders, and switchers in one patch, and then move to a nearby patch. Suppose that you know taxon A is a grower in the first patch, but that the second patch has a heavier or more variable herbivore load. What would your expectations be about A's sister taxa in the new patch? Different expectations correspond to different ways of understanding an adaptationist thesis about flexibility.

We can think that history delimits the scope of an adaptationist hypothesis. Historical adaptationists will expect that defenders in the second patch will be most closely related to defenders in the first patch; growers in the second, to growers in the first. Your best between-patch predictor will be history; the relatedness of your two taxa. But within a sister group, environment predicts difference. In infected environments, growers defend more; in relatively herbivore-free environments, defenders grow more. In varied environments, each switch more. So the adaptationist hypothesis is right, if nested within lineage, and seen as explaining variance within the lineage. This is a history-nested conception of an adaptationist hypothesis about defence and plasticity.

A stronger version of adaptationism is to think that adaptation overrides history. On that view, the best predictor of whether A's sister will grow, defend or switch is the herbivore load of the environment. Whether A defends carries information about whether A's sister defends, but only because and to the extent that the two species inherit similar niches in similar environments. Sister groups usually are similar, but their similarity is largely a product of the similarity in their environments; their relatedness does not carry much extra information. Adaptationist hypotheses become more historical the higher they are nested in an evolutionary tree. Thus Povinelli's adaptationist hypothesis is nested in the primate lineage. He expects primate minds to evolve a self-concept as they become housed in large and arboreal bodies. A more history over-riding version of adaptationism would push the evolutionary prediction deeper into the tree, and hence covering more diverse organisms. Tree kangaroos are large arboreal mammals. Hence their environment is complex in the same way. A bolder and less historically constrained adaptationism would predict the evolution of a self-conception in the arboreal macropod mind. So the environmental complexity hypothesis is really a family of adaptationist hypotheses depending on the extent to which its defender thinks selection over-rides history. I think more historical versions of adaptationism are both independently plausible and strong enough to sustain an environmental complexity hypothesis about the function of cognition.

A second issue in integrating the environmental complexity hypothesis with phylogeny is a serious worry about whether adaptive plasticity, or even cognitively mediated adaptive plasticity, is a single trait. For the models Godfrey-Smith discusses are necessarily simple with a very encompassing criterion of adaptive flexibility. He discusses this issue, defending his broad sense of adaptive flexibility:

> The environmental complexity thesis will be understood . . . as an adaptationist hypothesis about a core set of capacities that nervous systems make possible and which are displayed by a large range of animals.
>
> This is a large set of assumptions to make. I am not trying to damn the principle at the outset though. "Camouflage" and "swimming" can be regarded as single, very widespread traits which can be realised in diverse structural properties. It is not hopeless to develop a general theory of the place of swimming in nature. . . . Cognition might be similar; it might be a single trait with certain teleonomically central features, and with a small set of evolutionary factors bearing consistently upon it across different lineages. . . . (p. 23)

The more broadly we define swimming, the less likely it has an adaptive explanation, and the harder it will be to use phylogeny to test our adaptive hypothesis. The tree will become huge and poorly resolved. The decision to count our trait as present or absent will be less objective. Do squid swim? Like Godfrey-Smith, I have used inducible defence as my simple proxy for cognition. But my guess is that thinking about adaptive plasticity per se really does render our problem empirically intractable. So I think we need a less encompassing idea of cognitively mediated adaptive plasticity. Godfrey-Smith does suggest a sharper definition, distinguishing between first and second order plasticity.

> The organism which adjusts its behaviour to circumstances, but which does so in a rigidly pre-programmed way, has a first order property of complexity in its behaviour.

> Such an organism is inflexible in contrast to an organism which is able to modify its behaviour *profile* in the light of experience, an organism which modifies what behaviour it is that is produced in the presence of a given condition. . . . This is a second order property of complexity. (pp. 25–6)

However, second order complexity is satisfied trivially by an organism with substantial first order complexity. A fish that changes sex in response to size cues will change its behavioural profile too: other stimuli will induce new responses. I have defended the idea that cognitive representation requires more than complex response to a single specific proximal stimulus (Chapter 9). Ant hygiene, for example, is switched on by a specific proximal stimulus, oleic acid. Contrast the ant with the anti-predation responses of ravens, who recognise both different dangers, and the same danger through different cues. An organism that genuinely represents a given feature of its world must have several informational routes to that feature. There must be multiple channels between mind and world; organisms so equipped get behavioural feedback. So for me, the question becomes: in what lineages, and in what selection regimes, what kind of environmental complexity requires multi-channel tracking of the environment, and feedback guided response to that variability? This question might just be specific enough to be tractable while general enough to still be a question about the evolution of cognition.

IV REPRESENTATION, CORRESPONDENCE AND ADAPTEDNESS

Godfrey-Smith connects his basic explanatory picture to broader themes in epistemology. He combines the pragmatist idea that the point of cognition is the appropriate control of action with the reliabilist idea that the point of cognition is to accurately track the world. Accurate tracking is a means to appropriate behaviour. His discussion of these issues turns on the idea of a "fuel for success." Certain traits have value across a range of projects, and in many environments. Fuels for success *explain* success. Likability is a fuel for social success. The likeable are not invariably successful: sometimes they are hounded by obsessives or hamstrung by an inability to be brutal. But typically, whatever your social project, your expected success is greater if you are likeable. So being likeable will often be part of the causal explanation of some agent's success. Moreover, we cannot dispense with likability in explaining social success, for though it's based on more particular personality traits, it does not reduce to these. For folks are likeable in virtue of very different arrays of particular traits. Likability thus contrasts, on many views, with IQ. Does IQ measure a fuel for behavioural success – a generalised capacity that depends on a suite of more particular cognitive traits? Those who are sceptical think that IQ scores are like fragility. High IQ is not an explanation of behavioural success, but is just a specification of the propensity or disposition to succeed.

Godfrey-Smith compares the idea of representational accuracy to fitness or adaptedness in biology. The idea is that accurate representation is a "fuel for success," but that requires accuracy to be akin to likability rather than high IQ. The biological notions of fitness or adaptedness began life in biological thought as intuitive measures of an objective, overall feature of an organism that explains that organism's biological success. Grandma – the royal albatross that bred on the Dunedin peninsula from the mid 1930s to the early 1990s – was successful because she was highly adapted (or, equivalently, very fit). Grandma's adaptedness explained her success.

There has been a major shift in the official view of most biologists' conception of adaptedness. The official view of adaptedness (and, even more, of fitness) is that adaptedness is just the disposition to succeed. That view has been developed in response both to the difficulty in giving a clean account of adaptedness, and in response to the "tautology problem" for the theory of natural selection. The charge of tautology has been rebutted by distinguishing between actual and expected reproductive success; between realised fitness and the propensity to succeed. Fitness, or adaptedness, on this view is a reproductive propensity. On this conception of evolutionary theory we distinguish (i) the particular traits of an organism in its environment; jointly, these explain (ii) the organism's reproductive propensity, its expected fitness; and (iii) usually, that propensity approximates to an organism's actual reproductive success; its realised fitness. Adaptedness is linked to reproduction only in the way fragility is linked to breaking. Adaptedness, as a generalised fuel for success, has dropped out. An organism's propensity to succeed to degree X does not causally explain actual success of degree X. Rather, it's the organism's specific traits in its specific environment that does that.

Thus we have seen a shift in biological theory from the classical conception of the niche to the modern conception. The classical conception took niches to be "ways of life" in a community that can be identified independently of the organisms that fill them, and which can often be reidentified in a different community. On this view, we can make sense of two organism both being contenders for the same, independently identifiable slot in the ecosystem, and the idea that one contender fits that slot – is better adapted to it – than its rival. But this idea has drifted out, in favour of a notion of a niche that is specific to its occupier, so the idea of an "unfilled niche" becomes undefined (see Colwell 1992; Griesemer 1992; Hutchinson 1978; Schoener 1989).

Godfrey-Smith argues that we can synthesise reliability conceptions of epistemology with pragmatist conceptions, but only if mind-world correspondence is a general fuel for success. It can be so only if the fate of fitness does not overtake correspondence. The facts which constitute mind-world correspondence relations must be independent of the behavioural triumphs of the mind-owning organism, else accurate representations of the world will desiccate into mere dispositions to succeed in acting in the world. It seems plain that there are plenty of extent theories which are in danger of desiccating in this way. Any theory of meaning in which a "principle of charity" plays a central role, is a theory of meaning which will fail to generate a success-independent notion of correspondence. Godfrey-Smith argues that Millikan's ambitious version of teleosemantics goes the same way. I think this is right, and important enough to count as a constraint on naturalist theories of representation.

Godfrey-Smith explores teleofunctional theories of representation that are modest enough to give the notion of correspondence real bite. He accepts the dehydration of the idea of adaptedness but resists it for correspondence. I am inclined to resist for adaptedness as well. Accuracy as an aspect of adaptedness could explain success, if adaptedness could be rehydrated. In contemporary biology, fitness has been identified with expected reproductive success. But adaptedness still plays a double role, for there is still an explanatory need in contemporary biology for a generalised notion of fit. Thus:

> Biological structure, or some very significant portion of it, is understood as an adaptive response to environmental conditions. The primary mechanism for this adaptive response recognised today is natural selection on genetic variants . . . but this is not the only way to have an adaptationist

194

biology. Other mechanisms . . . can establish a relation of adaptation or
"fit" between organism and environment. (p. 32)

This is Godfrey-Smith speaking for the adaptationist, but in fact
evolutionary theorists often formulate their ideas in ways that presuppose we can make sense of a robust notion of adaptedness. Gould,
Lewontin, and Raup will not be suspected of being enthralled to adaptationist intuitions, so they can serve as examples. In Gould and
Lewontin's famous critique of adaptationism, they point out that selection can drive a change through a population which leaves each of
its members less well adapted than they were before (Gould and
Lewontin 1978). Their example is a fertility arms race in an insect population. Both Raup (1991) and Gould (1989), in thinking about macro-
evolution and mass extinction, distinguish between three possible
extinction rules. The "field of bullets" is random extinction: no property of the species predicts survival or failure. As far as I know, no-
one thinks that extinction in mass extinction events is random. Raup
formulates two alternatives. The "fair game" rule is formulated by
appeal to the notion of adaptedness: the better adapted a species, the
better its survival chances. Adaptedness is not expected fitness:
expected fitness is expected relative fitness, and compares the prospects
of organisms within a population. Adaptedness compares species with
one another. Gould and Raup do not think that mass extinction is governed by a "fair game" rule; this is one of the contrasts between mass
and background extinction. They believe mass extinction operates by
a "wanton" rule: features of a species predict its chances, but these features have nothing to do with adaptedness. If all this is right, in explaining the significance of mass extinctions in evolutionary history, we need
a robust notion of adaptedness. It helps explain extinction patterns in
normal times; its failure to do so in abnormal times is part of what
makes those times abnormal.

Moreover, if we lose an independent notion of adaptedness we lose
the central explanatory triumph of the Darwinian idea. For these
reasons, then, I would want to hang onto something like the classical
concept of the niche. Consider once more the breaking of a fragile
glass. The glass was fragile; it has a disposition to break. There is a specific explanation of its breaking; a causal story which locates the precise
mechanical forces that impacted on the glass structure, and details the
process whereby the crystals split. But there is also an intermediate
grain of analysis: an account of the structure and strength of the glass

195

which explains its vulnerability to impacts without specifying or explaining the particular breakage pattern that we observed.

The idea of adaptedness similarly strikes me as fitting into explanation at a similar grain. So think again about Grandma. The most specific description of Grandma is her full life history, and at this level, we lose the distinction between Grandma's actual and expected fitness. But we can abstract away from Grandma's full life history in various ways, by hanging onto elements of her life history that are "repeatable" or "systematic" in various ways but discarding those that are wholly idiosyncratic to Grandma. It seems quite likely that there will be no unique right abstraction. Grandma shares much of her history with other female royals nesting on the Dunedin peninsula, but some of it with royals anywhere; some of it with albatrosses anywhere; some of it with any fishing bird; some of it with any pelagic soaring bird. These abstractions away from Grandma's actual history give us a grip on an ideal life history of birds so characterised. The idea of the classic niche – or at least a relative of it – is the idea that there is a real point in some of these coarser, less fine-grained abstractions.[3] An externalist version of these ideas is the idea that when we abstract away from Grandma we will often be abstracting away from the internal features of Grandma's history, but not from the environmental features. Though that might be right, we do not have to accept this externalist idea to accept the idea that there is a useful coarse-grained abstraction where we can compare the ideal occupant of the niche to its various actual occupants.

V   FINALE

In sum, I agree with the general thrust of this book, but think there are unsolved problems both in the formulation and testing of the environmental complexity thesis. But this is a terrific book. It explores difficult and important intellectual terrain, and sheds light on every topic it touches.[4]

NOTES

1. One option would be to read this epistemically. An externalist might think that internal factors, though equally important, were intractably complex. This suggestion generates a local, modest and heuristic version of externalism. But I very much doubt externalists privilege the external over the internal only in

thinking that external factors are currently easier to study than internal ones.

2. I have been persuaded of this by Paul Griffiths. See his 1994 and 1996.

3. The idea that there is no single explanatorily relevant specification of the particular traits of the organism in its environment is akin to an idea Goode and Griffiths (1995) make in discussing the apparent ambiguity of biofunction. They point out that there are a number of correct, distinct specifications of the functions of the frog feeding mechanisms, as we move theoretical focus from neuroanatomy to natural history and ecology to population genetics.

4. Thanks to Fiona Cowie, David Hilbert, and Jim Woodward for their comments on earlier versions of this paper.

9

Basic Minds

I MINIMAL MENTALITY

Understanding representation in simpler cognitive systems than our own is important. It is often a good heuristic when interested in some issue to try first to understand its simple manifestations. Moreover, theories of animal representation are a direct test of theories of human representation: most obviously those that see a constitutive connection between language and representation. On plausible assumptions about the limitations of some animals' representational capacities, theories of animal representation test holist theories of representation, too. But the issue of animal mental representation is also important in its own right.

This chapter revolves around a distinctive phenomenon and two traditions. The phenomenon is the break – sharp or gradien – between representing and merely reactive organisms. The chapter focuses on the difference between having a minimum mind and no mind at all, and on some of the ways minimum minds can be upgraded. For much animal behaviour is the result of representation of the world, but not fully intentional representation of the world. For example, Ristau (1991b) shows that the piping plover's distraction displays are sensitive to environmental contingencies. She shows the direction of the display is appropriately linked to the position of the nest and the line of approach of the intruder and its intensity is linked to the perceived degree of threat both in closeness of approach and the current and past behaviour of the intruder. Yet piping plovers obviously are nothing much like a full intention system. Evans and Marler (1995) show similar restricted but genuine flexibility in chickens' responses to predators. They respond differently to aerial than terrestrial predators, and Evans and Marler show experimentally that these responses are not triggered by a single, simple easily characterised proximal stimulus (but see p. 263).

The two traditions are different ways of thinking about representation in the literature of cognitive ethology. Tinbergen, famously, distinguished the "four whys" of ecology: evolutionary history, current adaptive function, development and proximal causation. On one conception, representation is wholly within the domain of evolution and function. On these views, in attributing representational capacity to an organism, we commit ourselves to very little about the proximal mechanisms of behaviour. Millikan gives an account of representation which counts plants and bacteria amongst the informavores. Jerry Fodor also defends a wholly externalist theory of content. X is a cow representation in virtue of an X-cow relationship. Yet he wants to restrict the class of representation users to organisms whose mechanisms of behavioural control are suitably complex: paramecia do not represent their world (1986; 1990). In a recent paper, Allen and Hauser argue that Millikan's view of representation is at best incomplete on the grounds that it counts acacias as representing their world. On these views, then, representing one's world is something to do with the proximal mechanisms of behaviour. Dretske, in his 1988 restricts representation to learned representation. This is a developmental rather than proximal constraint, but nonetheless implies a different conception of why we talk of representation.

My aim, then, is to evaluate these two traditions of thinking about representation, and through that, explain the distinction between organisms that represent their environment, and those that merely react to it. On the basis of that distinction, I explore one example of upgrading basic minds. One major debate in primate psychology is over the alleged capacities of the Great Apes[1] but not monkeys to represent others' minds. I use the machinery developed to distinguish between representational and other biofunctions to motivate a tentative solution to this debate.

I begin with the simple end of the representation spectrum. When does representation begin? A fruitful set of examples are the complex behavioural patterns amongst the arthropods. The behavioural complexity of the bee's dance is well known. Distinct components of the dance co-vary with distance to a nectar source, direction to the source, and value of the source. There are many other examples. According to Holldobler and Wilson (1990, pp. 272–3), fire ants have an equally rich system: their odour trails carry information about direction and distance, and the number of ants using the trail carries information about the value of the food resource or of new nest sites. Desert isopods live in family groups densely – sometimes very densely – aggregated in suitable terrain. In natural circumstances, the foraging range of an isopod may include up to ten thousand other individuals. The family share a burrow – an absolutely essential resource – which is vigilantly and rigorously guarded from all others. No alien is ever admitted to a burrow. The burrow guard inspects and recognises kin in virtue of the "chemical badge" of each individual: a mix of individual and acquired traits that signals to the guard that the advancing isopod is kin. The badge in natural circumstances varies very precisely with family membership (Linsenmair 1987).

Communicative behaviour is another good avenue for seeing our way into the problems here. These are often seen as paradigm representations: hence Griffin's slogan (1991, p. 3) that communication is a "'window' on animal's thoughts." That slogan makes good sense if signals really are representations, and if signalling is information flow. The information that comes out of an animal's mind in communication is a sample – possibly a rich sample – of the information in it. So Ruth Millikan treats communicative signals as representations, even when the communicating organisms are not paradigms of intellectual power. Her paradigm representations (1989) are the beaver's warning splash and the honey bee dance. So what is claimed in claiming that a bee's dance represents the location and value of a food resource, a fire ant's trail the location to a new nest site, or a badge family membership? Is this claim just about the selective histories of these mechanisms or does it also depend on the proximate mechanisms of behavioural control? I begin with the proximally neutral conception of representation, and shall suggest that even those who are liberal about representation should be cautious about assuming that communication involves representation.

II SELECTIVE HISTORY AND PROXIMAL MECHANISM

Two liberals about representational capacities are Peter Godfrey-Smith and Ruth Millikan. Godfrey-Smith (1991; 1996) revives the Spencer-Dewey hypothesis that representational capacities evolve in response to environmental complexity. Thought is an adaptation to environmental complexity. This hypothesis is not obviously true: we may not have an independent measure of environmental complexity. Moreover, this view of the evolutionary history of cognitive mechanisms does not by itself distinguish representation from other biofunctions. Not every adaptation to complexity is a representation: size or digestive generalisation may be adaptations to complexity without being representations.

But Millikan is the archetype of the teleofunctional liberal. Representational explanation is biofunctional explanation.[2] But what kind of biofunction? She distinguishes between representation producers and representational consumers. Content is determined by use. We have representation when the user can do its job Normally only when the producer goes into a state that co-varies with a given environmental condition. The state then represents that condition. So she says "that the representation and the represented accord with one another, so, is a normal condition for proper functioning of the consumer device as it reacts to the representation" (Millikan 1989, pp. 286–7). For example, the mechanism in our respiratory system that results in panting detects the increase in CO_2 in our bloodstream, but it represents an oxygen shortage, for that shortage is the condition of panting fulfilling its proper function Normally. For something to be a representation, there must be something that takes it as a representation, and the content of the representation derives from how it is taken.

So the key idea is that a representation mediates between the producer and consumer in such a way that the consumer cannot fulfil its function Normally, unless the representation is veridical.[3] An ant food recruitment signal means "food" because the recruited ants cannot fulfil their function for the nest, namely of collecting food for the nest, unless the recruitment signal actually signals the existence and location of a sizeable food item. They may of course accidentally stumble on food when following a misinformed ant, but that is not a Normal fulfillment of their function. A similar story assigns content to beaver danger warnings and honey bee food dances.

In my 1990a, I defended a modest version of biosemantics. We should distinguish between a theory of what makes X a representation, and a theory of what X represents, given that it is a representation. I defended a selective history account of what X represents, given that X is a representation. I think part of the problem for Millikan's story is that it collapses these two issues into one. Millikan's theory of representation is beset by serious internal problems. In particular,

I think her flagship examples may be misdescribed. For her account of representation sees it as emerging from the interaction of co-operating devices. She may see a representational picture of beaver warnings and the like as unproblematic through a commitment to the classical ethological conception of communication as adapted for the co-ordination of two co-operating individuals. But it's at least arguable that picture does not best capture the biofunction of signals, hence will not lead to an appropriate evolutionary theory of signals. In their important 1984 paper, Krebs and Dawkins apply the paradigm of *The Extended Phenotype* to signals, and that leads to problems for Millikan's account.

The Krebs' and Dawkins' conception of communication poses a three-way problem for Millikan's conception of bee dance and beaver splashes. Once we see the divergent evolutionary interests between sig-naller and audience, the idea that communication functions as a flow of information is undercut, and with it the idea that signals are repre-sentations. Organisms often alter the behaviour of other organisms. Sometimes this action is direct physical action on the bodies or muscles of an organism. But many organisms make chemicals that act on their audiences' behavioural control mechanisms. Parasites sometimes alter the behaviours of their intermediate hosts in ways that facilitate their move to their definitive hosts. Chemical tools are used within species, not just between species. The suppression of ovulation in subordinate female naked mole-rats is chemically mediated. These chemicals merely function to suppress ovulation: I see no explanatory point in supposing they have the imperative content "Thou Shalt Not Ovulate!" Other operations of one animal on another can have a more benign appearance: food gifts from male to female spiders, or inducing egg for-mation by extra feeding. Sound can be a tool too, working on an animal's control system through its sensory channels: male canaries induce the development of the partners' ovaries by singing (Hinde 1970). Krebs' and Dawkins' picture of communication suggests that we assimilate the canary song to those of extra feeding and other merely chemical causes of behavioural change. If other signals are mere tools like the canary song, the result is an account of communication with no obvious role for representation.

Suppose, however, that we find principled reasons for resisting col-lapsing all communication onto mere manipulation. Another problem emerges. Signals do not just pass between co-operating devices. So if we do think that at least some signalling involves representation, it follows that representation is not best understood as involving the interaction of co-operating co-adapted devices. For example, gazelles "stott": they engage in a type of exaggerated and ritualised bounding when they sight an interested predator. If stotting means anything, it means something like "I see you, so don't even try to catch me." Now perhaps we should not think of stotting as communication, or as communication involving representation, at all. But if it fails to be a representation, it's not because gazelle and lion are not co-operating. Stotting has as good a claim to being a representation as, say, the ritualised play bows between one coyote and another that take place when one tries to solicit another into the mutually rewarding activity of play.

At first sight, this may seem a problem only for Millikan's choice of examples. But there is another problem that goes deeper. A number of friendly critics of teleofunctional theories of content have pointed out that such theories face indeterminacy problems, for there are more and less specific descriptions of selective regimes and functional outcomes (Agar 1993; Bekoff and Allen 1992; Goode and Griffiths 1995). A different problem for the specification of function arises from the perspective of the *Extended Phenotype*. Adaptations benefit genes, but not necessarily the genes of the adapted organism. Traits are compromises between replicators with at least partially divergent interests, hence their biofunction is less easy to specify. If so, then though signals are a phenotypic expression of the "signalling genes," so too, perhaps, is the behaviour of the audience. The modification in behaviour that is caused by the signal is part of the extended phenotype of the signalling genes. It can be expected to tend to benefit them. It may or may not be beneficial to the receiving organism and its genes. Many interactions are manipulative rather than or as well as mutualistic. Both signal and reception will be consequences of, and hence compromises between, both sets of genes. So when we think, along Millikan's lines, of the Proper Function of the dominant female's pheromones, we must ask: Proper Function for Whom?[4]

In Allen and Hauser's paper, the problem of assigning a function to rhesus monkeys' "food calls" emerges very clearly. The reasons to signal are very different from the reasons for response to signalling. For the respondent, it's clear that signals indicate food, and that is the value of responding to calls. It's less clear what the function of giving food calls is, since not calling is a high reward, high risk strategy. Monkeys usually find and punish those that fail to call. But a monkey that gets away with silence gets more. Once we notice that traits can be constructed by the joint operation of replicators whose replication interests are not identical, then assigning functions to those traits becomes much more complex. Moreover, if the perspective of *The Extended Phenotype* is right, it need not just be behavioural traits that are evolutionary compromises between different sets of replicators. This moves the problem of assigning function inside the organism.

Millikan is alive to the danger of an excessively liberal account of representation. She thinks that is one of the problems of indication theories of content. Chameleon skin colour co-varies with background, as does a weasel's fur colour over a year, but neither represent the colour of the animal's background. But Millikan's account is also a very liberal one. It counts ants and plants as representational systems. Acid detectors in the ant represent dead nestmates, for that co-variation is the Normal condition for hygienic removal to fulfil its proper function. Plants' daylength detectors represent the approach of spring, for those mechanisms that begin spring growth fulfil their proper function only when the triggering condition co-varies with seasonal change.

One of the points of this chapter is to try to push beyond a clash of intuitions over the liberalism that Millikan recognises and intends. However, I think her account is even more liberal than she intends. It might make any organism a representing system. For all will have some mechanisms tuned to features of their environment on which others depend. Consider, for example nuptial plumage of male birds. These mediate the interaction between male and female. They have the bio-function of attracting mates: perhaps by advertising parasite freedom, good genes or sexy sons. The female can breed successfully only if nuptial plumes veridically represent an acceptable mate. I think it's far from obvious that breeding plumes are *representational communications* between the birds: that breeding plumes represent the state of affairs of being a healthy mate even in the same sense that a bee dance represents the location of a good nectar source.

We might bite the bullet on nuptial plumes. But it will turn out that saliva represents food. The production of saliva in the mouth adapts the digestion system to the receipt of food. So it mediates between our mechanisms of ingestion and digestion: our digestive systems can only fulfil their function normally, if saliva is correlated with food. Yet surely this is a non-representational mechanism. I think the relations between many merely physiological mechanisms will meet Millikan's

conditions. The cross-section of my leg bone will represent my interaction with the gravitational field, for those aspects of my morphology
which depend on it for their structural support can only work Normally
if the cross-section is appropriate to the weight it bears.

In a recent paper, Colin Allen and Marc Hauser attempt to repair
the liberalism of indicator semantics and biosemantics by distinguishing between "strong" and "weak" senses of information. They believe
that the biofunctional and informational readings of signal meaning are
too weak for communication to count as a criterion of cognitive complexity. So they propose instead that:

> Signals of type S have strong information content for organism O just
> in case:
>
> (i)   tokens of S have weak information content C for O;
> (ii) other structures can have the weak information content for O that
>   a token S has occurred despite the absence of the conditions
>   described by C;
> (iii) other structures can have the weak information content for O that
>   no token of S has occurred despite the presence of the conditions
>   described by C. (1992, p. 89)

This condition seems too strong. But it is also too weak to demar-
cate the class of cognitively complex organisms. Any organism that can
be conditioned satisfies their conditions, and responsiveness to conditioning is very widespread in the animal kingdom. Consider the growth
and decay of the responses of a rat being conditioned to food by the
sound of a bell. The fact that this association can decay shows that the
rat is sensitive to the pair {bell, no food}. If the rat's rate of sponta-
neously checking the dispenser is sensitive to the world-head reliability of the bell, then the rat is sensitive to the pair {no bell, food}.
Nothing in any of this requires cognitive complexity of any striking
nature. There is, after all, no intrinsic reason why their criterion should
work. We have no reason to expect a simple strong relationship
between the type of phenomenon an organism is behaviourally sensitive to, and cognitive complexity.

I think Millikan's actual specification of biofunction lets in more
than anyone should intend. But she has and intends an account in
which the evolved systems of behaviour guidance of ants, trees and
chimps all count as representation processing devices. These systems
represent aspects of the organisms' environment. Fodor, Bennett, Allen
and Hauser, and others prefer an account of representation in which

"reflex-like" organisms do not count as informavores. So let's first characterise the control systems that liberals do and chauvinists do not think of as representation users. I then sketch an argument for a less rather than more liberal account of representation.

In his 1983, Dennett emphasized "performance specifications," characterising different control systems in terms of their different capacities. More recently he has argued that we use the notion of representation to capture "real patterns" exemplified in the behaviour of some systems. The problem, however, is to say which patterns require representational notions to capture them. Digestive systems exemplify real though noisy patterns. Fodor (1986) and Bennett (1991) have both tried out the idea that representation enables an organism to act, and action is selective response to non-natural kinds. The "real pattern" is a capacity to respond to non-natural kinds, however this capacity is physically realized. This idea puts more strain on the notion of a natural kind that it can bear. Desert isopods can distinguish their relatives through chemical cues, but being a relative of isopod number 47,012 is not a natural kind.

An alternative is to think of the flow of control. On this conception some organisms have a "straight-through" flow, responding to stimuli without contribution from internal storage or feedback. The behavioural mechanisms which produce "vacuum" behaviours – for instance, phantom egg rolling in greyleg geese – look good candidates for behaviours produced without the benefit of feedback loops or input from other sensory modalities. In his 1960, Tinbergen reports elegant experiments which reveal the herring gull's vulnerability to "superstimuli"; they brood oversize dummy eggs in preference to their own, and this might also show simplified control structures. Tinbergen's work on bird behaviour is full of examples of complex behaviours that are triggered by very simple triggers. Less surprisingly, the same is true of the arthropods. There is a single and simple physical stimulus for nest hygiene, and nest hygiene itself is simple removal. The stimulus can be characterised physically rather than informationally. The behaviour takes place whether appropriate or not, just so long as the simple initiating condition is present. Ants remove dead nest mates when they detect oleic acid the decaying ant releases. Thus a live though acidic ant will be expelled, struggle how she will.

Simple mechanisms result in a fixed though perhaps highly structured response to some stereotypical stimuli. The arthropods provide many examples of organisms capable of a few quite complex behaviours, but capable *only* of those behaviours, and with very little capacity to alter their behaviours when circumstances change. The inputs from these behaviours are "transduced"; there is no internal information contributing to the registration of a stimulus, so a given input produces the same response.

III REPRESENTATIONAL EXPLANATION

I think considerations about the explanatory point of appeal to representations support the chauvinist denial that organisms whose behaviour is controlled only by simple mechanisms represent their world. One version of a standard appeal to representation in explanation contrasts the diversity of internal organisations of the mind with the unity of the distal states to which those organisations respond. The internal mechanisms through which we avoid fire, flood and famine may be very different. Yet the behaviours produced are the same, and have the same cause. Despite differences in neural housekeeping, we represent the world in the same way and that is why we behave the same way.

An intuitively useful way of running this idea derives from a distinction drawn by Frank Jackson and Philip Pettit (see for example their 1992). They distinguish between **robust-process** and **actual-sequence** explanations.[5] Distinct explanations of the same event can both be important, for they can convey distinct breeds of modal information. In earlier work I have used the origins of World War I to illustrate this idea. This time, I will illustrate it through a more significant conflict, Australia's victory over England in the 1974–5 cricket tests. One explanation of this victory would walk us through a play-by-play description of the tests, detailing each dismissal, run by run and out by out. An alternative would appeal to the strengths and weaknesses of the opposing sides: in particular, Australia's strengths in fast bowling and fielding. These two explanations do not conflict, and each is of value. The play-by-play explanation is an actual-sequence explanation, for it identifies *the particular possible world* that we inhabit. But if it is true that Australia in that series was much the stronger side, we could know the precise sequence of plays without knowing something very important. Namely, had Australia not won that way, they would have won in another and similar way, with the fast bowlers taking most of the wickets, often to spectacular behind the wicket catches. A robust-process explanation *compares our world to others*. It identifies an important class of counter-factuals. It identifies the feature characteristic of the worlds in which Australia triumphed in the series.

When representations are causally relevant, they will be relevant as part of robust-process explanations of behaviour. For example, in two fascinating and convincing books, Franz de Waal charts the shifting balance of power in the Amsterdam chimp colony, and helps himself to richly representational description of chimp mental life on the way. Those representational descriptions are warranted, on this conception, if they are part of a robust-process explanation of, for example. Yereon's long struggles to maintain his position. What tracks worlds in which Yereon maintains his central role in the Amsterdam chimp colony? Let proximal stimuli and internal organisation vary as they may, Yereon's worlds will be all – or at least only – those worlds in which he represents and hence responds to nascent alliance between Nikki and Luit. Hence we best explain the patterns in Yeron's behaviour – the alternations of threat and conciliation through which he maintains his position – by attributing to him that representational capacity.

I once argued (Sterelny 1990b; see also Allen 1992) that there was a role for representational notions in comparative psychology, even the comparative psychology of simple organisms. Animals exhibit terrific variation in the structural and computational details of their control mechanisms. Bats and owls hunt the same flying insects at night; possums, bees and honeyeaters forage for nectar; many quite different organisms have enemies in common. Despite differences in internal machinery, their *Umwelts* – their lived worlds – overlap. That overlap explains their behavioural similarities. I am no longer confident of the strength of this argument. In those cases, like the ants, where we do have a genuine biofunction with different realisations we can recognise the *design* of signals without thereby being committed to the idea that signals represent anything. Though ants differ greatly, the group is unified by both phylogeny and ecology. It is much less obvious that there is a theory of warning behaviour in general. I very much doubt that we have either single cause or single effect in the warning behaviours of arthropods, birds and mammals. The greater diversity in the developmental and proximal mechanisms of behavioural control there is in the group on which we focus, the less likely it is that there is anything that they really share, anything that is a genuine convergence.

Some robust proximal explanations of behaviour require us to identify an organism's represented world. Development points in the same
direction. Complex behaviours of even simple organisms involve a
developmental trajectory. Since nests may be genetically diverse, "nest
smells" are not predictable. Pupa do not emerge with a capacity to
recognise nestmates. So virtually all organisms exhibit some form of
developmental plasticity. Nevertheless, not all plasticity is alike. Desert
isopods learn to recognise nestlings and new instars in a critical period
when these hatch or shed their old exoskeletons in the communal
burrow. Very likely, this depends on a very specific process.

Consider, in contrast, brushtail possums in New Zealand. These
recognise and learn to use a wide variety of foodstuffs. Many initial
conditions and proximal histories converge on similar capacities. A
robust explanations of their learning history will be representational.
A feature of the possum's environment best tracks the worlds in
which we get that trajectory. When a very specific proximal stimulus
is necessary, then it seems unlikely that a distal property will best track
the worlds in which we get that trajectory. Adult vervets fear martial
eagles. When they see them, their warning and avoidance behaviours
are quite distinctive (Cheney and Seyfarth 1990). If there are a
variety of routes through which eagle recognition and appropriate
responses to them are constructed, then we should explain this as learning about eagles and how to avoid them. But if the vervet behavioural
changes are cued by a few highly specific stimuli, there is nothing about
vervet behaviour or change in behaviour that is robustly tracked by
a feature of the world which we can take the organism to come to
represent.

So there is something right about the idea that developmental
plasticity comes in importantly different varieties, and that only some
learning is learning to represent the world. Dretske was onto something in suggesting that there are no innate representations. But despite
Dretske's use of it, I doubt that the notion of innateness helps demar-
cate the development of representational from other control systems.
The notion itself is foggy, conflating at least the ideas of phylogenetic
history, developmental rigidity and genetic control. One notion of
innateness does seem to capture something like inflexibility: a capacity is innate when its development is initiated by a minimum and
precise triggering experience. A classic example is that of goslings
imprinting on Lorenz. These do seem marks of acquisition processes
for which a robust-process explanation need not rest on a feature of

the world which the organism is learning to recognise and respond to. But some "poverty of stimulus" arguments are quite different. Some depend on the capacity of the organism to respond to complex, subtle and degraded information that does not wear its message on its face. The organism could only use that information, the idea runs, if it is cognitively pre-equipped for it. If that is right, there is no conflict between the idea that a developmental process is under the control of innate mechanisms and the idea that it consists in the acquisition of a representational capacity.

A consideration of the point of representation pushes us away from the idea that organisms with simple control systems represent their world. Consider a nest in which an ant has died. Under what circumstances, actual or counter-factual, will that ant be removed? The ant hygiene worlds are not worlds in which ants construct, through various proximal routes, a representation of a dead colleague. Rather, the worlds are tracked by a specific proximal stimulus: ants react to a *particular* decay product. Once initiated, that reaction involves diversity of bodily motions. How an ant removes a corpse, and where it is removed to, depends on the contingencies of individual and nest geography. That is why the ethological description in terms of proximal stimulation and the function of behaviour is robust. Had not an ant removed a corpse then and that way, she or another would have removed it soon and similarly. So a description of the precise sequence of biochemical and neural events would miss something, just as a play-by-play close-up of the cricket series does. So Jackson and Pettit's distinction cuts across those of Tinbergen. There is no one "why" of the proximate causes of behaviour. An actual-sequence explanation of (say) chimp behaviour details the transition from stimulus to response, and a robust-process explanation turns on what is represented.

So what tracks the possible worlds in which piping plovers feign a broken wing? The flexibility and fine tuning of the display suggests that the worlds are tracked by a representational property: the display worlds are worlds in which the object of the display is a threat to the nestlings. But with reflex-like organisms, the idea runs, what is robust is proximal stimulus not distal source. What tracks the possible worlds in which isopod number 47,002 is freely admitted to the family nest? Though kinship and badge recognition vary well together, the robust property is the badge, not what the badge is a badge of. Isopods have ingenious mechanisms for kin recognition. But they do not have an equivalent of our perceptual constancy mechanisms. Dretske (1981) pointed out our constancy mechanisms give us multiple routes to stable features of our external environment. That is why our visual systems represent an object's shape, not a retinal image. Since isopods have nothing like this, the property that tracks isopod response is proximal stimulus, not the objective property of the world that characterises the selective regime in which the isopod's response mechanism evolved.

Scepticism about isopod representation is not, of course, scepticism about the biological function of the badge. No doubt, that is kin recognition. But just as we could describe the biofunction of ant warning signals without using representational notions, we can do the same for the badge. It functions to admit kin and kin only to the family nest. Indeed, we can use Ruth Millikan's useful distinction between remote and proximal biofunctions to elaborate on them. She applies that distinction to the hoverfly: the proximal function of one of the hoverfly's visual specialisations is (in her view) to locate small moving black dots on its retina; its remote function is to represent the presence of a potential mate (1990). Similarly, the proximal function of the isopod badge is to match a stored chemical sequence; the remote function is to admit only kin to the nest.

So I incline to the view that simple organisms do not represent their world, rather than represent their world crudely. So what then are *representational* biofunctions? They are internal states which do not just have as an effect the detection of features of the environment; they have that effect by design. That internal state is there because, often enough, tokens of the same type enabled its ancestors to respond to that feature in ways that advantaged them by comparison with their less responsive conspecifics. So what is it to *represent* a state of the environment, as distinct from merely reacting by design to a proximal stimulus? It is for there to be sufficient variety in proximal routes, and sufficient stability of distal sources, for the organism's adaptive reaction to the environmental feature to be robust.

IV   PRIMATE PSYCHOLOGISTS

In my view, then, we can exploit the notion of a robust explanation to distinguish representation from other internal states that function to guide behaviour. I think the same conceptual tool is useful in identifying not just whether a state is a representation but also what that state represents. One standard worry about teleofunctional theories of content is that selective histories can be specified in a number of ways (O'Hara 1993). How many times need servals predate vervet ancestors for us to take servals as well as leopards to be part of the selective regime that shaped vervet "leopard" calls? In Chapter 6 I recruit the idea of robust explanation to characterise selective regimes. Adaptationist explanation makes a bold empirical bet. Sober and others have characterised it as the bet that other evolutionary forces have been unimportant in the evolution of an adaptation (Sober 1993). In taking the function of ants' formic acid to be defence, we take chance and history to be unimportant in the evolution of that trait. I argue that this mischaracterises the relations between selection and other evolutionary mechanisms: The adaptationist thesis is not that other elements are unimportant, but that the selective regime is robust. Had the variation available to selection, or the developmental or phylogenetic constraints been different, so too would be the evolutionary trajectory of the ant lineage. But if the selective hypothesis is right, so long as we do not vary the selective regime, all these differing trajectories will share a common feature. Adaptationist explanations are appropriate when but only when there is a robust counter-factual feature to be captured, and we capture it when we say what all the ant defence worlds share. So the counter-factual dependence of vervet calling on serval predation is what matters, not its frequency: "leopard" calls mean "leopard or serval" only if the predation by both cats is common to the calling worlds. That is possible: serval predation though rare might push predation to a threshold at which warning calls evolve.

Let's apply this machinery to a central debate in the contemporary primate literature: the idea that chimps and perhaps other close relatives have a theory of mind. The idea is that these primates are aware of themselves, and they are also aware that their fellows represent the world. This second-order representation of other's minds is rich enough to enable them to manipulate other's behaviour by manipulating their minds.

Some of the case histories are very striking. Let me run a few examples to give a sense of the intentional temptations here. Austin and Sherman are two chimpanzees in Savage-Rumbaugh's ape language project. Sherman dominated Austin but was afraid of the dark. Austin was not and would leave their joint cage at night, make strange noises (tapping in pipes and windows and the like) then come rushing back into their cage hair bristling. Instead of bossing Austin around Sherman would rush up and embrace for reassurance (Jolly 1988). Some experimental results push us in the same direction. Premack and Woodruff, for example, attempted to test the capacity of chimps to learn a connection between seeing and knowing. After pretrial familiarisation with opaque containers, four chimps were allowed to watch two trainers, as these in turn watched one of two of the containers being baited with food. Only one trainer, however, was in a position to see which container had the food; the chimps were in a position which would have allowed them to see which trainer that was. The chimps were then allowed to pick a trainer to seek advice from. The right trainer – the one that could see what happened – was always right; by prearrangement, the blindside trainer was always wrong. Two of the four chimps mastered the experiment, at least to the extent of a bias towards picking the right trainer and usually following the right trainer's advice (Premack 1988). A key component in the idea that some primates have a "theory of mind" is the social intelligence hypothesis (Byrne 1993; Byrne 1995; Byrne and Whiten 1988, 1992; Whiten and Byrne 1988). The idea is that the evolution of primate intelligence has been driven by the demands of a complex social life. In such a life fitness depends to a very great degree on an appropriate blend of co-operation and competition. Invariant rules of thumb work poorly in social environments. For interaction in social environments is strategic; how others act will depend on your actions and their expectations about your actions. The evolutionary idea, even if right, leaves a good deal open. It is compatible with two views of cognitive architecture. Evolution in a social context may have fuelled an increase in general intellectual and learning capacities. Alternatively, it may have driven the evolution of a specialised capacity to deal with social problems. Moreover, and most importantly for our purposes, the hypothesis has no very direct implications about primate psychologists. Social intelligence requires learning about and representing *the observable social world*: who is aggressive, who can be trusted, who is effective as a coalition partner. These capacities may not require apes to have a theory of mind. The problem that has worried many in this field is that primate knowledge of their social world may be awareness of behavioural conditionals, not mental structures. A solution to this problem requires us to see that the distinction between awareness of behavioural conditions and that of intentional structure is too stark. We should distinguish four hypotheses about the nature of primate mental representation. We can take the Great Apes to have:

i. The capacity to represent the social organization of, and the relations within their social group – The Weberian Primate Hypothesis.
ii. We can take them to have (only) the capacity to represent a relatively fixed range of behavioural conditionals true of their social partners – The Skinnerian Primate Hypothesis.
iii. We can take them to have the capacity to represent the multitracked behavioural dispositions true of their social partners – The Ryleian Primate Hypothesis.
iv. We can take them to have the capacity to represent the complex of motivational and informational states which explain their social partners' behavioural dispositions, and hence, for example, to be able to predict their social partners' behaviours in a wide range of non-stereotypical situations – The Austenian Primate Hypothesis.[6]

There is good evidence that the Weberian Primate Hypothesis is true. But its truth is compatible with any of the ideas represented by (ii) through (iv). The distinction between (ii) and (iii) depends mostly on learning history: as we discover that primates can anticipate conditionals for which they have no specific learning history, we should shift from the Skinnerian to the Ryleian conception of primate capacity. Our conception of the distinction between (iii) and (iv) will depend on general questions in the philosophy of mind. On Dennett's most recent views (Dennett 1991a) metarepresentation is just a threshold of complexity of computational processes involving the social and behavioural relations of self and others. On that view no single functional ability is essential to metarepresentation, or constitutive of it. The distinction between (iii) and (iv) would be one of degree only.[7] A devoted defender of the language of thought (Fodor comes to mind) would take metarepresentation to involve a specific computational mechanism; a particular recursion in the language of thought. Hence there would be a principled distinction between the Ryleian and Austenian hypotheses, however experimentally intractable it was. Either way, the distinction is an important one.

The functional hypothesis that the adaptive value of intelligence is to manage social complexity is compatible with a variety of ideas about the overall architecture of the mind and its representational capacity. It may also be compatible with more than one developmental mechanism. One of the most articulate and vigorous sceptics of the attribution of rich representational capacity to non-human primates is Celia Heyes (Heyes 1993a and 1993b; 1994a; 1994b; Heyes and Dickinson 1990). Amongst much else, she argues that many social abilities may be the result of associative mechanisms. For instance, female baboons have been observed to steal meat only from relaxed males and, indeed, to induce this relaxation by grooming. Heyes suggests association may well be sufficient to establish these and similar skills. So she contrasts two conceptions of the baboon. On the one view, the baboon associates relaxation through grooming with the safety of theft; on the other it knows that it is safe to steal when and only when the male is relaxed. But if association is a learning mechanism, it's by no means clear that these are rival views: one is an idea about what the baboon knows; the other, how the baboon has come by this information.[8]

In thinking about primate representational powers, methodological issues are blended with conceptual ones. The central methodological issue is the inference from behaviour to representational capacity. The problem arises for both demoting and promoting hypotheses. The evidence supporting promoting hypotheses – the idea that chimps do have a theory of mind – is typically subject to reinterpretation, with varying degrees of strain, in terms of instrumental knowledge or even conditioning. Similarly, demoting hypotheses can be challenged: our behaviour is far from a perfect reflection of our intentional structures, let alone a perfect reflection of the intentional structures we ought to have: false belief, superstition and weakness of the will intervene. If other primates represent the minds of others, sometimes they will misrepresent them and even when they do not, their behaviour will not perfectly reflect what they know and want. But in practice most of the worry has been about promoting hypotheses: what behaviours enable us to take primates to represent the mental worlds of their conspecifics rather than have mere instrumental knowledge. Cheney and Seyfarth (1991) suggest that though monkeys are sensitive[9] to the behaviour of their conspecifics, they show little sensitivity to what their conspecifics know. In experiments with Japanese macaques, for example, they show that the rates of alarm and food calls are not sensitive to whether the audience is aware of the threats or opportunities. Macaques show not much evidence of a theory of mind.

Heyes thinks that similar considerations apply to the Great Apes. She argues, in my view correctly, that many capacities taken to show a theory of mind can be given a more downbeat reading. Chimps are sometimes alleged to have a theory of not just others' minds but their own: to have introspective self-awareness. One oft-cited item of evidence for this is Gallup's mirror experiments. These purport to show chimps use mirrors to inspect their own bodies, and hence have introspective self-awareness: a capacity to represent their own representations. Even granted the effect is real, they show no such thing. In some sense any organism able to use perceptual input to navigate itself through an environment must be able to represent their own body location. The added sophistication in using a mirror for guidance is not metarepresentation but the ability to use unusual input. Chimps thus may have fewer hard-wired constraints on a sensory input that can be used.

"Perspective-taking" experiments have more potential to persuade. In these, the experimental set-up is supposed to confront a primate with a task whose solution requires the primate to distinguish its perspective on the world from that of another agent. But even these are subject to re-interpretation. In a demanding version of these experiments, chimps watched containers being baited with food in the presence of two trainers. Both were present while a container was filled with food, but Guesser had a paper bag over his head. Both trainers then indicated a container: only the Knower was a reliable guide to the right one. Chimps were allowed to select one container, and two of them managed to learn to (usually) take the one indicated by Knower. The successful chimps were taken to have succeeded in representing the visual world of the trainers (Povinelli et al. 1990). But even here, the chimp may need to know no more than "if he has a bag over his head, take no notice of his guesses." The successful chimps may not know why Guesser's guesses are not accurate. Similarly, Jessie, Premack's chimp who understood enough to remove the blindfold from a trainer rather than simply drag the trainer to the locked container may have understood only the connection between eyes and behaviour, not eyes and the mind. (Premack 1988, p. 165).

Deception also illustrates the problems of inferring from behaviour to capacity. Some of these cases give the Austenian hypothesis terrific psychological pull. A very striking example is that of Nikki and Luit suppressing their signs of anxiety, and concealing their faces until they did, while struggling for control.[10] Engaging tales of subordinate chimps concealing their erections strike a chord. But perhaps all subordinates know is that erections cause attacks. Colin Allen (pers com) made the interesting suggestion that we use pornography to induce erections, and hence test for a theory of mind. If the young chimps have a theory of mind, they should realise that when there are no females around the display of pornography-induced erections will not result in aggression. But even here there is a conjunctive conditioning hypothesis: if I show my erection when there are females around I will be beaten up. When are such suggestions ad hoc? Moreover, we do not want to make the conditions of metarepresentation too hard to pass: we can be conditioned in ways that are dissonant with our represented world. A case in point are phobias that the agent knows to be irrational. So Danny might realise he has nothing to fear, but still cover his erection out of a conditioned aversion: a chimp superstition about the ill-luck of showing his erection to his betters.

The cautionary remarks of Heyes are right: many of the deception anecdotes and experiments may show only that primates are good behaviourists rather than good psychologists. But there are a range of increasingly complex appreciations of what others will do. These begin from limited behavioural contingences: if I show my erection he will bite me. A more sophisticated understanding involves a range of conditionals rather than a single one: if I show my erection Mike will be angry, with anger involving quite multi-tracked dispositions. The Ryleian hypothesis, once we distinguish it from the Skinnerian and Austenian ideas, is very plausible. Jolly is surely right in thinking chimp and other primates are aware of others' emotional states, and hence are aware at least of quite complex behavioural contingencies. De Waal's work on primate politics seems to show at least this level of sophistication. Whiten and Byrne point out that some of the most plausible examples of deception involve manipulations of attention, so the Ryleian chimp may be aware of some of the consequences of the fixation of attention and of others' perceptual perspectives. These are all intermediates between mere knowledge of behavioural contingences and paradigmatic representation of others' motivational or cognitive states.

Chimps and other primates represent rather than merely react to their social world. The work of Goodall, de Waal, Jolly, and others shows that their response to their circumstances is not under the control of a simple proximal stimulus. They represent their social world in part by representing certain aspects of members of their social group. Let's suppose that selection in social evolution has shaped Darwinian Algorithms in chimps specialised for social life. What is common and peculiar to these selective regimes? What features of their associates have the animals in the chimp lineage responded to over evolutionary time?[11] Birds represent predators, because bird response is tracked best by features of predators, not by features of the proximal stimulus. Similarly chimp behaviour is tuned to dispositions rather than particular stimulus/response conditionals. The perspective-taking experiments do not demonstrate a chimp theory of mind, but they do show that chimps do not need a specific learning history for each circumstance/behaviour pair that they understand. In my view, their representations are Ryleian. Consider Mike, the chimp that invented banging on kerosene tins to make his charges and displays more effective. The interactions around Mike's change in status show both sophistication and its lack. If all Mike knew were a fairly fixed set of behavioural counter-factuals, Mike would have no expectations about the use of kerosene tin banging in display. Equally, his associates would have had no expectations about Mike's jealousy about his new devices. This they would need to learn separately. But equally, the whole story shows a lack of sophistication too. Banging a tin does not make Mike any more powerful. His associates were either superstitious or rather easily tricked. As Dennett has occasionally mentioned, genuinely Austenian primates ought to be puzzled by the behaviour of the trainers. Trainers are weird. Why don't they pull their own blindfold off, or discard the paper bag from their head? The behaviour of the trainers – Trainer Right and Trainer Wrong – in the perspective experiments is pretty odd: standing an equi-distance from the bins with the hidden food, not moving until a string is pulled, then simply handing over food. What could they be doing? What are they up to? An Austenian primate ought to be deeply suspicious about these regimes. So I think our best guess is that the selective regime is one in which chimps have acquired a sensitivity to, and the capacity to respond to, complex dispositions of their fellows.[12]

NOTES

1. I apologise for this biologically unsound term: the only group that includes orangs, gorillas, and the two chimps also includes us.

2. See especially her 1984, 1989, and 1990; I will draw especially on her 1989, as it is the most accessible of her accounts of her ideas.
3. A second condition is that a signal be part of a representational system; it has a "syntax." But it is easy to meet this condition, since the time and place of the signal – as in the warning slap – can satisfy it.
4. See Chapter 3 for an attempt to extend the notion of biofunction to cases where traits are influenced by replicators with divergent interests.
5. They use the terms "comparative" and "contrastive," but this terminology invites confusion, for "contrastive" has been pre-empted by Dretske.
6. I refer of course to the famous English primatologist, Jane Austen.
7. Byrne and Whiten, building on the work of Leslie, have argued for a constitutive connection between metarepresentation and other abilities. They think that the capacities for "symbolic play" and for attributing thoughts to others are not distinct functional capacities but the same capacity differently described. I am very sceptical. For one thing, the category of symbolic play itself strikes me as subject to over-interpretation. Some of the primate anecdotes have terrific appeal: Vicky pretending to pull a toy, and pretending to unsnag it gets a real grip on our intentional imaginations. Leslie (and following him Byrne and Whiten) argue that a chimp pretending that a banana is a telephone involves both representation of a banana and a metarepresentation. In my view, pretend play involves two representations – straight and "as if" – of a single object, not a representation of a banana, and a representation of "banana" as a telephone. Moreover, I see no clean difference between these cases and standard mammalian exuberance. The kitten in chasing leaves, say, has dual representations of the leaf, for it does not try to eat the leaf. It does not *just* represent a leaf as prey. And kittens often chase tiny/imaginary gnats – their play does not have to be triggered by a real object.
8. We can drive a wedge between the two ideas if we have a very tight definition of association. So we might define learning as the result of an associative mechanism if there is stimulus generalisation only to physically similar proximal stimuli. Unless there is a specific trigger (as there is in the feeding responses of gull chicks and the like) in the social scene, it is surely unlikely that association narrowly defined will play a major role in primate interactions, as the number, shape, orientation and relations between the interactants will vary considerably from social setting to social setting. So unless there is a specific invariant feature across these social settings (as there might be in male-male displays: distinctive teeth baring and the like) the social actors need to be sensitive to socio-functional similarities from situation to situation (the alpha male is restless, or aggressive, or distracted) rather than proximate similarities.
9. Sometimes. Vervets continue to give alarm calls long after everyone has responded and fled to safety. So on the assumption that alarm calls mean "leopard" rather than "I'm scared" even sensitivity to behaviour is rough and ready.
10. De Waal, quoted in Jolly (1991), pp. 239–40.

11. I have presented these arguments in the context of a teleosemantic view of representation. I think, however, that if chimps have their representations of each other in virtue of ontogenetic causation, then the ideas here can be written as a claim about the robustness of causal processes in ontogeny.
12. Thanks to Marc Bekoff and Jack Copeland for their comments on an earlier draft of this chapter.

10

Intentional Agency and
the Metarepresentation Hypothesis

> ... [I] may turn out that a profoundly new psychology evolved
> quite recently in the history of the primate order. In its wake it
> may have left two fundamentally distinct groups of life: those who
> know that a mental world exists and those that do not.
>
> (Povinelli 1993, p. 494)

I INTRODUCTION

Virtually all organisms have devices which function to modify their
states in response to a change in the environment. But not all living
creatures represent aspects of their world. For some organisms that
respond even in quite complex ways to their environment nevertheless
have rigid behaviour guidance systems. The arthropods, for example,
include many organisms capable of quite complex behaviours, but
capable *only* of those behaviours, and with very little capacity to alter
their behaviours when circumstances change. Their behaviours are
driven by very specific proximal stimuli. For instance, the desert isopod
has a marvellously intricate and accurate system for detecting kin, for
only kin are admitted to the communal nest to escape the heat of the
day. That mechanism depends on a chemical badge that the isopod
acquires when it hatches within the nest, and which is recognised by
the nest guard. The guard's behaviour is under the control of a very
specific proximal stimulus. It has no other access, no other perceptual
or informational route, to the distal property of kinship. So the most
robust explanation of the guard's behaviour appeals to the badge the
guard detects. Kinship and badge vary well together. But while isopods
have ingenious mechanisms for kin recognition, they do not have an

equivalent of our perceptual constancy mechanisms. The property that tracks isopod response is proximal stimulus, not a distal feature of the world that is the adaptive rationale for the response mechanism. So a nest guard does not represent an approaching isopod as its kin.

Contrast the isopod with the New Caledonian crow, *Corvus moneduloides*, a bird which makes a number of different twig tools which it uses to extract invertebrates from holes and other inaccessible nooks and crannies. The flexibility and fine tuning of the manufacture, use and storage of these tools suggests that these crows represent features both of the tools and of their use. There is no specific stimulus that unlocks tool making. So I count the crow but not the isopod as a representational system. (See Chapter 9 for further defence of this distinction.) Nonetheless, it seems unlikely that the New Caledonian crow is much like a fully intentional system. I do not think crows have beliefs and desires. While the evidence is not in, I would bet that there are considerable restrictions on their representational powers. I think of the crow as a paradigm representational system, but it's also a paradigm subintentional system. We are paradigm intentional systems, or intentional agents. What is the difference?

Dennett (1995) is sceptical about questions of this form. In his view, the evolved nature of the mind undercuts the intuition that there must be an essential difference between real intentional systems and lesser cognitive engines. He may be right. Representational systems may differ from one another only in degree, or along many dimensions. Some organisms may be able to represent only features of their immediate environment. Others will be less stimulus bound, able in varying ways to represent the absent.[1] Organisms' representational engines may be more or less domain bound. Cheney and Seyfarth document both vervet monkeys; relatively sophisticated understanding of one another's social relations, and their crude capacities for assessing evidence about predators (Cheney and Seyfarth, 1990). Organisms vary in their learning mechanisms. Some learn the features of their world to which they should attend from their social group. Others may have the ability to dodge trial-and-error learning by copying skills from their fellows. We can learn from our conspecifics' representations, not just their acts. Intentional systems may not form a natural kind, or anything like a natural kind, within the class of representational systems.

Intentional systems may not form a natural kind, but evolutionary theory alone does not tell us so. Moreover, our cognitive capacities are striking in at least the following respects:

1. Our representational capacities are neither stimulus bound nor restricted to a few predetermined domains.
2. Our cognitive development relies on advanced social learning. It has still not been decisively demonstrated that other animals can use their peers' behaviour as a model. We can learn not just by imitation but from others' representations.
3. We are strategic agents. We are very adept at predicting the behaviour of other agents, even in novel circumstances and even when their behaviour varies from its previous pattern. We can, for example, anticipate opponents learning from their mistakes.

Evolutionary biologists sometimes use the concept of a grade to characterise a distinctive adaptive complex. Eusociality – "true sociality" – is such a grade. It is characterised by co-operative foraging; the division of reproduction within the group; the emergence of sterile workers; morphologically differentiated castes; and a good deal more. One important and interesting question within evolutionary theory focuses on identifying eusociality and the conditions under which it evolves. Intentional agency, too, might be seen as a grade, a distinctive adaptive complex, and that raises in turn the problem of identifying it and the conditions under which it might evolve. One reason for interest in other organisms is to test the extent to which these representational, developmental and inferential capacities form a package deal. A broader comparative perspective might show a fundamental factor: the cognitive keystone for intentional agency. Moreover, there is a candidate for that key factor.

The upbeat literature on primate cognition, and the associated comparative literature from developmental psychology is largely focused on capacities with two characteristics.

1. These capacities are thought to be present, though in varying degrees, in adult humans, developing humans, and the great apes.
2. They are thought to require metarepresentational capacities.

The usual examples are Mirror Self-Recognition, Pretend Play, Imitation, and the "Theory Theory" version of Social Intelligence Hypotheses. The idea is that the capacity to metarepresent – to form representations of representations – underwrites a new and distinctive set of abilities, and constitutes the key cognitive adaptation of our lineage. Metarepresentation is critical to advanced social learning, mediated by imitation and "symbolic" play. It is critical to negotiating complex social environments. So the idea on centre stage is that there is indeed a distinctive seam in the class of representational systems, namely, the metarepresentational systems. This seam divides those organisms capable of "second order" intentionality from the others (Dennett 1983). Those that accept primate metarepresentation are those that give intentionally rich explanations of primate behaviour (for example, de Waal; more cautiously, Povinelli, Premack, Byrne and Whiten).[2] Though they are not always as explicit as Povinelli (1993), they take metarepresentation to be the distinctive feature of intentional systems. In this Chapter, I discuss two exemplars of metarepresentational analyses, imitation and the "theory-theory" of social intelligence. Starting out as a friend of the metarepresentational view, I have turned into a sceptic.

II SOCIAL LEARNING, IMITATION AND METAREPRESENTATION

Many animals do not come pre-programmed with food preferences, foraging styles, and, more generally, an appreciation of the opportunities and dangers of their environment. Their social experience is important to their development. Oystercatchers acquire from their parents one of the two approved oyster-opening methods. Naive monkeys develop fear of snakes by exposure to the fearful responses of adults. Chimps and gorillas often need to learn quite complex processing techniques to exploit some of their food sources. So socially mediated learning plays a role in the ontogeny of much animal behaviour.

The prevailing view, however is that most of these skills are not acquired through "true imitation." Learning can be socially mediated in a number of ways.

a. As a by-product of their own daily routines, the adults in a group may transform the flow of experience to an infant. Such "environmental structuring" may markedly affect that infant's learning. In effect, the group determines the infant's immediate environment. Standard trial-and-error learning then produces typical adult behaviour.

b. "Social priming" or "stimulus enhancement" may direct the attention of the infant to events and activities about which the infant should learn. A mother eating a banana focuses her infant's attention on bananas, whose possibilities it is then more apt to explore. This accelerates trial-and-error learning. The infant differentially explores its immediate environment.

c. The activities of group members may not just direct the attention of an infant onto an activity. The infant may become aware of the outcome of those activities, and seek in its own behaviour similar outcomes. Its social experience teaches it about certain possibilities, but not how to make those possibilities actual. A young chimp may notice that adults can get nuts by somehow banging away at them, and thus explore various ways of placing and hitting nuts, intending to somehow open them, not having noticed how the adults turned the trick.

d. In "true imitation" an infant takes as its model not just outcome but means, and consequently represents both its own and adult means, using discrepancy to improve its own performance.

True imitation is taken to require great cognitive sophistication. To fix ideas, let's continue with the nut-opening example, which is one of the candidates for true imitation amongst chimps. A mimic sees a model opening nuts with a hammer and anvil. The mimic wants to open nuts, and it transforms a representation of the model cracking them as seen from a spectator's vantage point into a representation of nut cracking seen from the point of view of the agent itself. The mimic is then able to compare its attempts to open nuts with this stored and transformed representation of successful opening in improving its own attempts. Byrne, Whiten, Ham, and others think this involves metarepresentation and perhaps even something approaching a theory of mind.

> To imitate in the visual mode involves B copying an action pattern of A's that was originally organized from A's point of view. . . . It is necessarily a different pattern from B's point of view, yet it has then to be re-represented in its original organizational form so as to be performed from B's point of view. The expression "re-represented" seems unavoidable and is used advisedly: it translates as second-order representation or metarepresentation. . . . To put the idea graphically; we might say that B has to get the program for the behaviour out of A's head: in other words, to engage in a type of mindreading. The hypothesis predicts that, as acts to be imitated become more complex, so it will be difficult to achieve imitation when the viewpoints of model and imitator differ, as opposed to B watching over A's shoulder. (Whiten and Ham 1992, p. 271)

I think true imitation is cognitively sophisticated. But I think this account of its sophistication is mistaken. First, the metarepresentational hypothesis conflates the transformation of a representation with a metarepresentation. Some animals can form cognitive maps of their territories from procedural information. Having made some trips around their territory, they can then find direct, previously unused routes between two points. They must transform their representations, but need not construct metarepresentations. Similarly, if I form a representation of what it would be like for me to use a Macintosh Mouse from a representation of another person using one, that is a transformation of one representation into another, not the representation of a representation. So talk of true imitation *by itself* showing metarepresentation is wrong.

Second, even the idea that imitation involves transition between points of view may well be mistaken. It makes a tacit assumption about the mimic's representation of the model's behaviour. A transformation between points of view is needed only if the mimic represents something like the model's motor pattern as seen from some specific point of view, and turns it into a representation of a motor pattern from another. But that is not the only possibility. Consider the skills involved in using a hammer and anvil to open tough nuts. If the mimic represents the model's behaviour functionally – pick up a rock in the grasping hand; hold the nut facing away, place it on a smooth hard surface – there is no need to transform between points of view. If in seeing nut cracking, the young chimp constructs something like a recipe, no point of view is involved.

Byrne himself makes a similar point in introducing the idea of a *behavioural programme*. Behavioural routines may have an overall functional organisation, rather than consisting of a mere chain of independent behavioural atoms. If some skills depend on behavioural programmes, imitation can involve copying that programme rather than a specific motor pattern. Gorilla food preparation illustrates this idea of a programme. Since they often eat thistles and other rather awkward plants, gorillas often need to do a good deal of manual processing of their food before they can eat it. If young gorillas acquire this ability through imitation, they may be employing any one (or some combination) of:

1. the logical structure is copied (program level imitation), but the motor act sequences used in achieving each stage are not;

2. some motor-act sequences are copied in detail (impersonation), but an idiosyncratic logical organization is used;
3. the entire sequence is copied, which could involve either or both mechanisms. (Byrne 1995, p. 68)

Byrne's distinction here is well taken, but it is fatal to his claim that imitation involves perspective taking. It suggests instead that the true cognitive skill involved in imitation – or at least programme level imitation – is functional analysis. On this view, imitation might still be significant for it involves an important capacity for abstraction. But it has no special connection with metarepresentation.

I strongly suspect that one reason why true imitation has been taken to be a sign of intelligence is its perceived distribution. The friends of the primate mind suggest that imitation is restricted to creatures we think of as intelligent on other grounds: the great ape clade, and – just possibly – dolphins. True imitation is thus correlated with intelligence. This correlation, and the significance attached to it, throws great weight on our capacity to empirically distinguish it from other forms of social learning. Yet this correlation is a factoid rather than a fact. Firstly, it depends on treating birds' mimicry as a distinct phenomenon.[3] Second, the optimistic assessment of ape capacity largely rests on anecdotal reports, especially of humanized apes. There are many such reports, but there is little experimental vindication of them. Yet experimental evidence is particularly important for those who emphasize the cognitive distinction between imitation and other social learning.

Evidence that would distinguish true imitation from stimulus enhancement is hard to find in natural settings. Thus Byrne (1995, p. 75) puts considerable weight on observational and anecdotal evidence about parrots, apes, and dolphins, because the reported behaviours were not species typical. For that reason, they are not likely to be a consequence of the animal's normal maturation, nor be an expression of a skill already in the animal's repertoire. So the behaviours could result only from imitation or trial-and-error learning. If the experimenter can rule out trial-and-error learning, only imitation remains. When studying animals in the field, there are more possibilities to be ruled out. It will rarely be possible to exclude the possibility of prior learning, for we rarely know the full learning history of field-studied animals. One option is to find or devise problems which have multiple solutions, to see whether the naive subjects tend to copy not just output but also procedure. So we could experimentally test for imitation versus stimulus enhancement by designing food filled boxes that can be taken to pieces equally well, though with difficulty,[4] starting in more than one direction.[5] There are unlikely to be natural experiments with equivalent solutions. Even where there are alternative ways of solving problems that are posed by an animal's natural environment, it is quite likely that one solution will be easier for the animals to find, given their abilities, than the others. So even if the population does use a relatively homogenous method, that similarity may be due to convergence on their best method. So, to establish the importance of imitation, we need to show more than similarities in method, more even than the similarities in the fine grain of motor patterns that Byrne (1995, p. 67) accepts as evidence for imitation. We need to establish arbitrary similarity: similarity in methods not functionally required by the goal. This is a tough task for field studies, hence controlled experimental results are of some significance.[6] Yet the evidence that true imitation is both well developed in, and nearly restricted to, the great ape clade is not compelling.[7]

A third reason for taking imitation to be cognitively significant is a hypothesis about its role in human development. Gopnik and Meltzoff argue for the cognitive and developmental importance of imitation (Meltzoff and Gopnik 1993). They think imitation develops at birth, and that it is critical in developing a "theory of mind." The idea that imitation is connected with the awareness of oneself and others is very plausible. But Gopnik's and Meltzoff's ideas depend on an extremely rich interpretation of very simple behaviours of very young infants. These interpretations invite sceptical downsizing, especially by those from the primate world who have learned from bitter experience how carefully one needs to guard against Clever Hans and other experimental artefacts. The evidence for true imitation looks decidedly thin. There is no novel behaviour, there is no complex behaviour, there is no functional structure to the behaviour. We cannot even draw a clear distinction between what is done and how it is done. Byrne (1995, p. 59) suggests that these apparently imitative acts are just the result of behavioural contagion. They are like an epidemic of yawns at a dinner party. The fact that the infant's behaviour is both contingent on and similar to an adult's behaviour does not show that the infant needs to know that any similarity exists.

Byrne is perhaps a shade too sceptical. Gopnik's and Meltzoff's evidence suggests problems for the ultra-deflationary account.

a. They claim that the infant's imitation improves. Infants appear to use the adult's act as a target which provides feedback for their own performance.
b. The greater the range of adult behaviours the young infant could match, the less plausible it would be to think that it is simply following a series of innate independent rules of the form: "If adult does X, do Y." This would be very plausible if babies just smiled in response to smiling, or cried in response to crying. But they can do more than that – they can poke out their tongues and do a few other movements which seem to have no obvious adaptive point. It is unlikely that these are pre-wired behavioural routines with a particular trigger.
c. Without some representation of similarity, it is hard to see how the infant could recognize that it is being imitated. Yet there is some, albeit weak, evidence that babies can do this. For there is experimental evidence that they attend more to those mimicking them.[8]

So perhaps the contagion thesis is too sceptical. Moreover, I suspect that Gopnik and Meltzoff are right in thinking that imitation is very important in the ontogeny of human social skills and social learning. The limited imitative capacities of autistic children; the early appearance of something like imitation in human infants; its ubiquity in young children; and the existence of elements of imitation – monitoring of gaze and attention – in our close kin are all striking. But while the skill of imitating complex behaviours is cognitively sophisticated, it is not a metarepresentational capacity, but rather a capacity to analyse a behavioural routine into its functional components. Moreover, though abstraction and functional analysis are important, it is by no means obvious that they mark a cognitive divide of any kind. For all I can see, they might come in smoothly varying degrees.

III STRATEGIC AGENCY AND THE "THEORY-THEORY"

I claimed in Section 1 that humans are strategic agents, very adept at predicting the behaviour of others. Strategic agency may very well be fundamental to human cognition. For one suggestion has been that cognitive evolution in the great ape branch of the primate lineage has been driven by a cognitive "arms race." In that lineage, fitness depends heavily on social skills, on negotiating the right mix of coalition, co-operation, and conflict. Those social skills, in turn, are rightly seen as cognitively demanding (Byrne and Whiten, 1988). At first sight, the adaptive significance of strategic agency seems to underwrite the elevation of metarepresentation to the cognitive keystone of full intentionality. For the "theory-theory" is one candidate explanation of strategic agency, and it is a metarepresentation hypothesis in spades. According to the "theory-theory," a strategic agent does not only have metarepresentational capacities; such an agent has an organized theory-like representation of the thoughts and cognitive dynamics of other agents, and uses that representation to predict and explain their behaviour. We have within us a very reliable Prediction Engine, and that Engine works by having within it a theory of the human psychological representations and their connection to behaviour from which it generates predictions.

It is worth noting at the outset that "theory" in this literature is used in a very robust way. So, for example, in their review of the developmental literature, Astington and Gopnik characterize "theory" as:

> Theories are abstract; they postulate theoretical entities that are far removed from immediate experience or evidence. They are coherent: there are complex law-like interrelationships between these theoretical entities. Theories allow one to generalise, to explain, to predict. They have a complicated relation to the evidence . . . while theories coherently organise many different types of evidence, they are still relatively domain specific. (Astington and Gopnik 1991, p. 17)

Some defenders of the theory-theory think that this commitment to the explicit theoretical character of these internal representations is gratuitous. Michael Devitt, for example, thinks that a theory-theorist can hold that the theory of others is only implicitly represented in the agent's Prediction Engine (Devitt 1996, pp. 79–80). A serious problem for this agnostic suggestion is that we lose our grip on the distinction between behaving on the basis of an implicit theory of agency, and behaving as if one had a theory of agency. Spiders, in building their webs, behave as if they had a grip on the theory of physical lattices, but I do not think they have within their control structures any such theory. Yet if a theory can be implicit and yet still guide action, we may have no basis on which to reject a theory-theory of spider web building. So I shall focus on Astington and Gopnik's robust version of the theory-theory.

Despite the metarepresentational character of the theory-theory, I am sceptical of a fundamental link between intentional agency, strategic-agency, and metarepresentation, mediated through the "theory-theory." The standard versions of the "theory-theory" are modular hypotheses, and these are problematic both in themselves, and in tension with the connection between strategic and intentional agency. The more one thinks our capacities to deal with our social environment are the result of a domain-specific cognitive adaptation, the less convincing is the link between strategic and intentional agency. An intentional agent must be capable of navigating cognitively over a reasonably broad range of domains. Thinking of the theory-theory as a module demotes strategic agency from its status in explaining intentional agency.

Moreover, there is an alternative and I think more plausible view of strategic agency, simulation theory. Simulation theory requires metarepresentational capacities, but posits much less fancy ones than the theory-theory. The essential idea is that an agent uses his/her own machinery to predict/explain the behaviour of other agents. So the agent need not have a theory of that machinery, or even represent it in any way. Instead, the agent needs some means of feeding "simulated" beliefs and/or desires into his/her own decision-making machinery, some way of taking the output "off-line," and, of course, some way of representing or otherwise using that output (see Goldman 1993; Goldman 1995; Gordon 1995). So metarepresentation enters simulation theory most obviously in representing output off-line.

I shall begin by discussing modular versions of the "theory-theory." Perhaps the theory-theory has to be modular. Our understanding of other agents as agents is automatic, mandatory, and develops early in human ontogeny. It is universal in normal humans. Moreover, if it's a theory, it's an unconscious theory. Appropriately prompted, we can extract banal, hedged decision rules: if agent A desires that p, believes that q-ing will result in p, then A will q. But it's hard to believe that the power, flexibility and precision of social navigation rests on these. If not, our theory of other agents – if it is a theory – is both deeply unconscious and efficiently used, just as is our modular mastery of our language. These are all marks of a modular capacity.

Yet on the other hand strategic interaction is not a good case for an encapsulated mechanism. Reading the intentions and beliefs of other agents contrasts with language parsing, species recognition, or object recognition in vision. These are domains in which we might expect procedures relying on restricted data to work well. Both speaker and hearer want to achieve uptake, even if in a deeper sense each wants to mislead the other. Object recognition in vision can rely on assumptions about the nature of the visual environment which are general and stable. There is no reason to suppose agents in general want to be intentionally transparent. There is no reason to suppose that outward signs of intentional states will have remained stable over long periods so that evolution can build specific recognition mechanisms. There is a vivid contrast between one human calling another's bluff, and – for instance – assessment by leopard seal males of their rivals. For a seal's size is both a critical objective basis of success and hard to fake. In contrast, the outward signs of particular intentional states need not be hard to fake or conceal. So it is not very plausible to suppose that there are reliable symptoms of intentional states to which an encapsulated system could be tuned.

Moreover, the infamous poverty of the stimulus argument does not extend in any clear way to strategic agency. The central plank of linguistic nativism is the claim that the information needed to master a language is not present in the primary linguistic data to which children are exposed (Pinker 1994). But there are two ways to read a poverty of the stimulus argument. Nativists read it as follows. There is insufficient usable information in their primary linguistic experience for a child-as-theorist to construct from scratch a theory of their language, so they do not construct it from scratch. The child has an informational head start. As many have pointed out, there is a second message one can take from the same premise. The poverty of the stimulus shows the child achieves its competence by some other means than by constructing a theory of their language. That second response has never been very impressive in responding to Chomsky and company, for want of an alternative. But the "theory-theory" is not nearly so well developed as a theory of agency, and in simulation theory there is an alternative.

Finally, we have the problem of fitting everyday folk psychology into a modular version of the theory-theory. On the MIT view, folk grammar has little connection with universal grammar. If the theory in the theory-theory module has nothing significant in common with folk psychology, we have the problem of explaining how it generates predictions conceptualized in those terms. For we can predict not just behaviour but intentional states. If the Prediction Engine runs Folk Psychology™ 6.1, we have an unexplained and aberrant isomorphism between the theory built into the module, and Granny's maunderings about the mind. So defenders of the theory-theory are in a somewhat awkward position whatever choice is made on modularity.

IV THE DEVELOPMENT OF STRATEGIC AGENCY

The development of strategic agency has often been seen as undercutting simulation theory, and supporting the "theory-theory." I disagree. I think simulation theory is well placed to explain the gradual but uneven processes of development. We can imagine the capacity to simulate developing as agents learn (1) how to simulate both beliefs and desires; (2) how to simulate a wider range of inputs, and hence develop the capacity to deal with agents increasingly unlike themselves now; (3) how to exploit different elements of their own machinery by feeding into them simulated inputs. Perhaps first they use decision-making machinery, then inferential machinery, as the agent learns to predict not just what others will do but also what they will think. Greg Currie has recently defended the idea that visual imagination depends on a simulated use of visual machinery (Currie 1995). So it is not difficult to see how capacities could arrive piecemeal, as the child's capacities to simulate different inputs, and take different outputs off-line, develop.

One controversial topic in developmental psychology is play. In my view, simulation theory fits it rather well. Play is a controversial topic, for there is supposed to be an important distinction between standard mammalian play of the kind puppies and kittens engage in, and "symbolic" or "pretend" play, allegedly confined to humans and our closest relatives. Thus Vicky, the Gardener's young chimp, engages in "symbolic" play when she pulls an imaginary toy train after her. I am somewhat sceptical of this distinction. Kittens, for example, routinely chase imaginary (or at least invisible) gnats, so I suspect Vicky's behaviour is seen as significant only because of pre-existent views of her cognitive powers. However, even if there is a distinctive category of pretend play, I am very sceptical indeed about the claim that it involves metarepresentation. Leslie has developed this idea. In his view, when, for example, a child uses a banana as a toy telephone the child both represents that banana and in some way represents their representation of that banana (see for example Leslie 1994; Leslie and German 1995). Toys vary in the degree to which they physically resemble their real counterparts. A child speaking into a banana shows an ability to use a toy which is physically distant from its real counterpart. In my view, simulation theory is well placed to explain examples of this type. We can think of play as training for loading into the action generator simulated beliefs and desires, and also, and perhaps especially, for taking outcomes of the action generator partially off-line. For pretend behaviour matches quite a lot, but not all, of the real behaviour.

In contrast, pretend play does not fit a metarepresentation hypothesis. Toy ducks are not representations of ducks. A banana, used as a toy, is not a representation of a telephone. A name is one thing, a surrogate another. Leslie's analysis of play as metarepresentation begins with his idea that the child cannot (in some "primary" way) represent her banana both with the thought *this is a banana* and with the thought *this is a telephone*. For then they would be involved in some kind of contradiction. So Leslie argues that the representation *this is a telephone* is "bracketed off" some way. It's "quarantined" from primary representation. Even if we accept that Leslie is right about this (and all that is obviously true is that the child does not straightforwardly believe that *this is a telephone*), at most this shows a representational transformation rather than metarepresentation. Jarrold suggests reconfiguring Leslie as proposing a new propositional attitude (Jarrold, Carruthers et al., 1994). The child pretends that this is a telephone. Just as I can without contradiction want this to be a glass of good red, while believing it to be a glass of vinegar, so too a child can pretend that her banana is a telephone while believing that this is a banana.

One problem with this revision of Leslie's suggestion is that young children recognize pretence in others. Three-year-olds do not just play, they seem to recognize play in others. Children of that age do recognize some mental states in others. But, intuitively, if pretending is properly understood as a distinctive propositional attitude towards some content, it is not a factive attitude, it is not a direct copy of their environment. So if young children understand that others have pretends-that attitudes towards contents, we might expect them to have the metarepresentational resources to pass false belief tests. Yet by and large they do not. More generally, this whole class of theories of play seems to me to miss the explanatory bus. The initial problem is to explain how children could represent their world by thinking both *this is a banana* and *this is a telephone*. One solution, the one we have been canvassing, is to say they have distinct attitudes towards those contents. Another is to claim that this formulation of the problem misdescribes the content of the child's thoughts. One content is *this is a banana*. The other is something like: *in the play context, I can use this as a telephone*. That is, it can *function* as a telephone. Like imitation, the cognitive sophistication of pretend play is that children develop functional and not just physical categorization of objects. This approach suggests how the representation of the banana as having contextually dependent telephone-like functional characteristics generates play behaviour. Inventing a new propositional attitude sheds no light on the connection between pretend representation and play behaviour.

I think development poses other awkward problems for the theory-theory. For how does the theory-theory itself develop in the maturing agent? The usual picture of development we take from the theory-theory is that a developmental sequence is a theoretical sequence. As the child becomes a strategic agent, she constructs increasingly more adequate theories of the inner life of other agents. We are strategic agents in virtue of internalizing a complex theory, and we have that capacity as the end result of a sequence of theory changes from cruder to better theories of mind. If so, we ought to be able to say what those theories might be like. Yet though developmental psychology has much to say about acquisition, it has found it difficult to develop a coherent theory of these precursor theories. Once again, there is a striking contrast with language. Derek Bickerton, to take one example, has given a quite detailed description of "protolanguage": a system explaining premature linguistic competence (Bickerton 1990). Let me give one example illustrating the problem of the precursor theory.

By 3, young children have made impressive but gappy progress towards strategic rationality. They have some elements in place: a hold on the pretence/reality distinction, and a grip on the distinction between different visual perspectives. So they know that if someone else is shielded by an opaque screen from what they can see, that person cannot see what they see. They understand the point of hiding things from view. Hence they seem to have some grip on the connection between perception and belief. But they have lots of trouble with deceptive appearance; for example, with sponges that are disguised as rocks, so looks hard while feeling soft. Three-year-olds attempt to resolve these by denying that appearances are deceptive, insisting that the disguised sponge looks like a sponge (Astington and Gopnik 1991,

p. 10). Famously, they fail an assortment of false belief tasks, even when prompted in various ways. Even though they do somewhat better on false desire than false belief tasks, considerable numbers fail them.

It is striking that it is very difficult to characterize just what model of the mind such children have. Gopnik and Meltzoff suggest that children think that minds are direct copies of the world. But that metaphor says nothing about a child's grip on desire: they do not think that all desires are satisfied. It is silent, as far as I can see, on the appearance/reality problem. If anything a copying picture might lead you to expect that perception dominates belief, rather than vice versa. It does not explain quirky partial success on false belief problems. Though children usually fail to predict on the basis of information about false belief, they can quite often explain on that basis: The puppet is looking in the box because she falsely believes a chocolate is in it. They can sometimes even predict behaviour when they have been told (and can later report on) the true location of an object rather than having seen it; perception seems to cause them to be in some way stimulus bound. Nor does it explain the child's success with perspective taking. A copy theory of the mind would predict failure on perspective taking tasks. The patterns of success and failure do not look theoretically integrated.

I should not overstate the strength of this objection. After all, generations of philosophers and psychologists have failed to reach agreement on a characterization of the adult theory of mind. Moreover, Gopnik and Wellman (1995) argue that the most intensively studied period of development is also a period of theoretical transition. The "copy theory" of mind is breaking down, and the child is attempting to patch it with various auxiliary hypotheses which in turn are deployed in rather ad hoc ways. Hence the appearance of messiness. Of course this could be true. No doubt three-year-old behaviour can be made compatible with a theory-theory. But it's hard to see the developmental capacities revealed by the studies of three- to four-year-olds as evidence for a theory-theory of strategic rationality.[9]

VI CONCLUSION: METAREPRESENTATION AND SIMULATION

It is time to stand back from the details. How do these issues bear on the metarepresentation hypothesis? Simulation connects to metarepresentation in two ways. First, if the simulation story is right, it down-plays the role of meta-representation in social intelligence. As the ability to simulate becomes more sophisticated, the range of pretend inputs the agent can feed into his or her own mechanisms increases, and the agent learns to feed simulated inputs through more of our cog-nitive engines. Metarepresentation will play a role somewhere in this process. For when we run inferential mechanisms off-line, we treat the results as predictions about what others will believe, and that predic-tion is a metarepresentation. But simulation could begin without inten-tional concepts. The output of a simulation need not be the ascription of a thought to another. It can be a behavioural prediction. The inputs will not typically be thoughts about thoughts, but will instead be rep-resentations of the world. The inputs will thus not *use* the concepts of belief, desire and other folk psychological concepts.

So far it seems that simple simulation could begin without the con-cepts of belief, desire and the like in place. What, however, of control; of selecting what inputs to feed into the action generator? At some point in the development of simulation, representation of the inten-tional states of other agents is an essential part of the front end of sim-ulation, of the construction of appropriate pretend inputs. However, I am not convinced that this is needed at the beginning of the process. Think of an early simulation. James watches Joe pull David's hair, and wonders what David will do. James feeds into his action generator a representation of that experience. This must involve a transformation of what he has seen, for the pretend input will be a representation to James having his hair pulled. That transformation might be something like those involved in perceptual perspective taking, when one agent understands what another will see from their point of view. So simula-tion must involve some capacities to transform representations, but I do not see that this transformation involves the apparatus of folk psychology.

So, even though metarepresentation is an important aspect of our strategic agency, simulation may be more developmentally funda-mental. It may be more evolutionarily basic as well. First, this idea gains some support from a suggestion of Goldman. He proposes (1995) that simulation is needed to explain conditional planning. In Goldman's eyes, conditional planning involves loading simulated inputs through the agent's decision mechanisms, and taking the out-put off-line. Goldman's idea finesses a potential "trajectory problem" for simulation theory. What use is a very basic simulator? The ability to simulate agents just like yourself now is unlikely to be a useful social tool. But if conditional planning does exploit simulation, it needs less fancy control mechanisms. Simulating other agents involves much more radical shifts from the agent's current state than does conditional planning. Hence the simpler control mechanisms of conditional planning could provide the foundations from which more complex ones evolve.

Second, simulation theory integrates well with the Machiavellian Intelligence hypothesis. On that view, the functional heart of human intelligence is social intelligence. Human cognitive evolution has been driven by an arms race amongst the human population. The biological fitness of our ancestors depended on co-operating with those who did not cheat you and avoiding co-operating with those who did. Judgements of belief, desire, and motivation were critical to one's prospects. The guess is that this is a feature of the great ape clade, though this particular adaptive transformation has been particularly intense in our case. Nonetheless, our close evolutionary kin show to a lesser extent many of these same cognitive adaptations. We have become intentional systems through evolution for social intelligence. The Machiavellian Intelligence hypothesis is at best a speculation, but it is a plausible speculation, because of the feedback mechanism built into it. It's very hard to judge complexity claims about the environment, but I agree with those who think we are smarter than we need to be to deal with the non-human world.

Now if all of this is right – needless to say, a big if – the mechanism that enables us to be strategic agents should be some mechanism selected because it boosts our social skills, but which has spin-off capacities in other fields. Simulation might well fit this bill better than a theory-theory. For example, if autistic children's failures are the result of inability to simulate inputs, then we can see how a mechanism selected for its advantage in one domain might eventually prove to have spin-offs in many. Autistic children are terrible at the "Tower of Hanoi" problem and its relatives, so even if the key adaptive advantage of simulation is enhancing social skills, its effects on capacity are not restricted to the social (see Currie 1996). It's much less obvious how the evolution of a theory-theory could generate intelligent capacities outside the domain for which it is selected. It's possible, of course, that Folk Psychology™ 6.1 requires such a fancy machine to run it that building that machine itself guarantees capacities outside the domain. But I do not see any way of supporting that conjecture.

In sum, I think the Machiavellian Intelligence hypothesis has some plausibility in the explanation of human intelligence, and hence of our being intentional systems. But if the explanation works, it does not depend on metarepresentation. If strategic agency is critical to our being intentional agents, metarepresentation is unlikely to be the key to the distinction between intentional and subintentional systems. Moreover, since simulating minds seem likely to differ from one another along a wide variety of dimensions, I am less optimistic than I once was that this is a distinction between natural kinds.[10]

NOTES

1. The unrefined and sluggish mind
   Of *homo javanensis*
   Could only treat of things concrete
   And present to the senses.
   (Quine 1953, 2nd ed 1961, p. 77)
   And while Quine may thus understate erectus' conceptual powers, this specification doubtless captures the limitations of a good many representational engines.
2. See for example: de Waal 1982; Byrne and Whiten 1988; Premack 1988; Povinelli 1993; Povinelli and Godfrey 1993; Byrne 1995; de Waal 1996.
3. The distinctness of birds might be defended on two grounds. Perhaps, in contrast to visual information, acoustic information does not require the transformation to accommodate point of view. This impression may however be an artefact of our perceptual limitations. More acoustically sensitive organisms might perceive this information as having directionality and point of view. Second, and I think more reasonably, perhaps the sound sequences birds learn to imitate have no organized structure. Lyrebirds and other avian mimics just chain together sounds. But these chains have no internal functional organization. So acoustic mimicry does not require abstraction. As usual, parrots turn out to be potential counter-examples. Even setting aside acoustic mimicry, they seem to have the ability to impersonate (Byrne 1995, p. 72).
4. The task has to be quite difficult, to give an animal a reason to learn it by imitation rather than trial and error.
5. This is the "two-action" test of Whiten and Ham 1992; see p. 254.
6. Boesch (1993) resists this conclusion. He argues that experiments disrupt natural ontogeny. So he argues for field studies of imitation looking for complex skills, especially if they were acquired early, and evidence of explicit teaching. In his view, hammer and anvil nut cracking satisfy these conditions. He argues that the skill is quite complex. Wooden hammers often need to be shaped, or stone ones found and their locations remembered. Apparently the positioning of the nut needs to be quite precise so that it is opened but not crushed, and it has to be hit in the right way. He argues as well that there is explicit teaching. Mothers provision infants with nuts and hammers. More

occasionally, mothers seem to teach the correct placement of the nut on the anvil or the limbs on the hammer by what seemed like explicit demonstration, i.e. very slow and deliberate action. It would be interesting to know if they monitored infants' attention, too. This whole example is clearly suggestive. But judgements of complexity are difficult to vindicate. It's certainly hard to show that these skills are so complex that socially facilitated trial-and-error learning could not explain their acquisition. Moreover, the evidence of teaching is clearest for stimulus enhancement, not true imitation. The mothers may be doing something sophisticated here, but teaching reduces the cognitive demands on the infant. That is the point of teaching – so no fancy capacities of the infant are on show here.

7. For a sceptical review of imitation in primates, see Galef 1992.

8. The evidence is weak, because it depends on the preferential looking time paradigm. This paradigm depends on an inference from the time spent looking at one of two individuals to the infant's representation of its focus of attention. This inference is fragile: an infant may recognize that there is something surprising, unusual or interesting about one of two matched adults (the one imitating it) without being able to represent what is surprising about that adult.

9. For more on both modularity and development, see Sterelny 1997.

10. Thanks to Fiona Cowie for her incisive comments on an earlier version of this paper, and of feedback from groups at Stanford, University of Maryland and Macquarie University.

11

Situated Agency and the Descent of Desire

This chapter is focused on the transition from *detection* to *representation*. Virtually all living creatures have mechanisms adapted to, and directing response to, specific features of their environment. No living thing is simple, but the simplest such creatures, bacteria, certainly have such adaptations. So, famously, *E. coli* have a gene which switches on in the presence of lactose, and whose product enables the bacterium to use that food source. So what distinguishes the capacity to detect and respond from the capacity to represent and act, and under what circumstances will the capacity to represent and act evolve from the capacity to detect and respond?

Godfrey-Smith defends the idea that environmental complexity selects for behavioural plasticity. If environments vary in ways relevant to the organism, then responding differently in different circumstances will pay better than a single fixed behavioural pattern. This is true, of course, only if the organism has some means of tracking variation in the environment and matching its behaviour appropriately to that variation. The world must be complex, but in ways the organism can map, if minds are to pay their way. For minds are not free. They require resources to build and maintain them. The adult human brain is only about 2 percent of total body mass, but consumes around 20 percent of total energy absorbed (Dunbar 1998). Furthermore, operating them has the cost of error, too. Trying to do the right thing at the right time risks doing it at the wrong time. In general, what ain't there cannot break, malfunction, or be misused (Godfrey-Smith 1996). Godfrey-Smith spends most of his time defending this idea for very simple forms of plastic response. One of his key examples is inducible

241

The Descent of Mind

defence by sea moss; animals not widely regarded as intellectual giants. But I shall suggest some ways of extending his idea from the simple detection mechanisms on which he focuses.

Mental representation has been the focal problem in philosophy of mind for the last twenty years. The orthodox view is that human behaviour in particular, and intelligent behaviour in general, is explained by explaining the capacity of the mind to generate, transform and use representations of the world. This theory, the representational theory of mind, is haunted by the notorious "frame problem." If we consider the complexity of real environments and the menu of actions available to agents, each of which has different potential effects on the environment, the problem of maintaining and using an accurate and relevant representation of the world seems intractable (Dennett 1984). So perhaps this is a misconceived approach to intelligence. Developmental systems theorists deny that the genotype is preformed information that guides development (Griffiths and Gray 1994; Oyama 1985). In a similar vein, there has been the development in the Artificial Life literature of anti-representational models of intelligent action. These models deny that intelligent behaviour must be guided by pre-existent belief and goal structures within the organism. Sometimes this line of thought suggests that the information to guide behaviour is constructed in behavioural interaction with the world, by organisms actively searching for specific, relevant cues. On these views, intelligence is the result of the interaction between organism and environment. Sometimes the idea is that information is stored in the environment, not just the brain, of the organism. Many spider mating rituals fit the interactionist idea rather well, for the action of each participant is contingent on signals from the other at each step. There is no obvious need to suppose that either participant has a stored template of the whole procedure if, instead, each stage completes the precondition for the next. Ant pheromone trails, on the other hand, obviously fit the picture of organisms storing information in the world rather than in themselves (Brooks 1991; Hendriks-Jansen 1996; for a critique see Kirsh 1996).

This conception of intelligent action without an internal representation of the agent's world is often called a theory of situated agency, to emphasise the importance of context in the control of action. Agents that act intelligently in the world without benefit of a prior representation are "situated agents." This theory of agency makes four central claims:

1. Behaviour can be partitioned into task-oriented skills. These are behavioural modules, each with distinct sensing/control requirements.
2. The behavioural repertoire of even complex creatures can be built from these modules, adding increments to a base of simple skills.
3. "Classical AI" has underestimated the information available in the environment. In the view of those sympathetic to antirepresenta-tional views, that extra information suffices to control these basic skills. These skills can be appropriately triggered and guided by local cues. "World models" are almost always unnecessary.
4. Organisms co-ordinate their behaviour through a built-in motiva-tional structure using information that the environment provides.

So as the organism moves through its environment, it interacts with it in ways that generate a variety of appropriate behaviours without it having to have and update a model of its world. At its most ambitious, the programme suggests that the "frame problem" is a pseudo-problem. It depends on mistaken presuppositions about intelligent action. But for the idea to be generally applicable, an organism must typically find itself in an environment which provides it with reliable cues. Moreover, these cues must be *local* cues. Situated agents escape the frame prob-lem by avoiding having any *overall model* of their world, even an overall representation of their immediate environment. For that is how they escape the problem of update. The behaviour of cue-driven organ-isms does not derive from the execution of stored templates nor is it otherwise controlled by information held in the organism.

Here we have the classic A-Life theme of emergence. Complex behaviour – the intelligent, adaptive behaviour of organisms in their environment – emerges from locally governed interactions between relatively simple components. A complexly behaving system need have neither complex parts nor central direction. A standard methodologi-cal moral emerges as well, in the idea that the *interaction* of the com-ponents determines system level behaviour. Hence we do not get much of a handle on what the system will be like by studying the components in isolation. The emergence of complex behaviour can be understood only through new models of scientific explanation (Sterelny 2000a; Burian and Richardson 1996; Clark 1996; Clark 1997; Hendriks-Jansen 1996).

This general methodological thesis links naturally with the tradition within ethology of scepticism about lab-based, experimental studies of behaviour. That tradition is still alive; for example Russon, in writing about the social expertise of primates, grumbles that "What expertise they actually do exploit this way is an empirical question. The relevant evidence comes from field observations, not laboratory experiments – the concern is how primates themselves use social learning, not how clever experimenters can induce them to use it" (Russon 1997, p. 183). If organisms are situated agents, storing some of their operational information in the environment, or developing it in interaction with very specific cues in their natural environment, caution about the experimental, manipulative approach makes sense. We would need to be very cautious in extrapolating from laboratory behaviour to natural behaviour and back again. In shifting an organism from its natural environment we may be extracting part of the cognitive system, not the whole cognitive system. Furthermore, interactions between the components, rather than the intrinsic features of the components themselves, are of most importance in generating system level behaviour. So studying a component in isolation will have limited utility for understanding the system. Equally, this picture vindicates scepticism about comparative psychology based on a few model organisms. The basic neural equipment – the basic internal mechanisms of control and learning – may well be fairly similar from organism to organism. But if the explanation of behaviour depends critically on the interaction between these intrinsic mechanisms and their environments, then neural similarity is no reason to expect overall similarity.

II THOUGHT IN A HOSTILE WORLD

What are the scope and limits of situated agency? When should we expect acting on local cues to be good enough? Just about all organisms, by design, track some features of their environment. I have previously argued that there is a very significant change when creatures become able to track their environment via more than one kind of proximal stimulus (Chapter 9). Organisms that can track features of their world in several ways represent rather than merely detect features of their environment. This capacity to track functionally relevant features of the environment in more than one way is required for behavioural capacities to be robust.[1] Arthropods often have beautifully ingenious ways of detecting relevant features of their environment, but they are often dependent on a single proximal cue. Thus the hygienic behaviour of ants and bees – their disposal of dead nestmates – depends on a single cue, the oleic acid decay produces. They have nothing equivalent to perceptual constancy mechanisms, mechanisms that would enable them to track the liveliness of their nestmates in different ways. Since ants recognise and react to one another by specific chemical and mechanical cues, parasites bearing no physical or other resemblance to the ants can invade and exploit their nests by mimicking the right specific signals. Thus, Holldobler and Wilson (1990) describe a number of beetle species that live in ant nests, persuade their hosts to feed them and even to tolerate their feeding upon the ants' larva by mimicking the ants' chemical signature and the food begging gestures (pp. 498–505). So though efficient in the right circumstances, their capacities are fragile.[2]

An organism that can track its environment only through a single, specific cue is very limited in its ability to use feedback to control and modulate its behaviour, for it is restricted to reliance on variation over time in that single cue. So if the cue can be misleading, or if there are time lags in response (as there will be with chemical cues and other physical signals whose transmission speed is low), feedback will be at best crude. Moreover, cue-bound organisms are unlikely to have behavioural abilities that are robust over a range of different environments. Environmental shifts are likely to disrupt situated cues. If animals have several channels to features of the world that are important to them (some of which may be mediated by memory), their capacities to act will be less subject to disruption through environmental instability. For example, an organism with a mental map of its territory will be less apt to get lost than one that simply follows a list of procedures ("right at the big tree, left at the wombat hole") cued by specific stimuli.

So cue-driven organisms will often struggle if their action plans depend on rapid feedback. More generally, they will struggle if ecologically relevant features of their environment – their functional world – map in complex, one-many ways onto their transducible world. Such organisms live in *informationally translucent environments*. If food, shelter, predators, mates, friends and foe map in complex ways onto physical signals they can detect, cue-driven organisms' behaviour will often misfire. Such organisms face the problem that many different sensory registrations form a single functional category, and similar physical signals may derive from very distinct functional sources.

But why and when would organisms find themselves in such environments? What evolutionary or selective histories might result in an environment being translucent? For translucence is a feature determined in part by the evolutionary history of the lineage, not just fundamental physical processes. "Nestmate" is a transparent property of the environment for many social insects. One reason is that organisms are often distributed through many different environments. The greater the variety of environments in which an organism might find itself, the less it will be able to rely on constant physical signals to identify food, shelter and its other needs. When a population is spread across many different niches, selection cannot predict the informational specifics of the organism's world. Koalas and echidnas may well be able to rely on cues – may well be situated agents – in recognising food. I doubt whether impala are. The relationship between the physical signals generalists can detect and the functional features they need to know about is likely to be complex. Hence the organism will need to represent its world. It will need multiple routes to the environmental features of interest to it. Organisms can be generalists not just through inhabiting a wide range of different environments but also through opportunistically exploiting a wide variety of resources in a single environment.

So ecologically more generalised organisms will be under some pressure to escape cue-driven behaviour. I think hostility also matters. Situated agency is a plausible picture of adaptive action only with respect to indifferent features of the environment. It is no accident that in this programme, A-Life models are of robots interacting with their physical surrounds; succeeding, for instance, in navigating their way around the walls of a room without having any overall representation of the room and its layout. Cue-driven behaviour offers a plausible picture of interaction with an indifferent physical environment. Thus we can conceive of, say, beaver dam repair as a sequence of skills driven by local cues. The sound of running water could initiate a random search along the inside of the dam walls; the feel of the current near a break could then induce a local search, then simple behavioural rules could guide action in dam repair. For this task the beaver need not have any overall model of the dam and its state. Equally, cues can be sufficient to drive behaviour in co-operative interaction, where each organism is trying to make its intentions as explicit as possible. Ants communicate with one another by producing local cues and mammals often have a single distinctive way of communicating the desire to play. But much animal behaviour takes place in a hostile world of predation and competition. Predation is not just a danger to life and limb; predation results in epistemic pollution. Prey, too, pollute the epistemic environment of their predators. Hiding, camouflage, and mimicry all complicate an animal's epistemic problems.

For cue-driven behaviour to be adaptive, the cue itself must be detectable and discriminable, though detecting a cue may require the organism to probe its environment. Further, there must be a stable cue-world relationship; that is, the organism's local environments must be homogeneous with respect to the cue-world relationship. What the cue tells the organism about its world needs to be fairly independent of what else is in the local scene. That is why the organism does not need to represent that local scene as a whole. Hostility imposes a cost of probing. It imposes a cost on animal action taken to disambiguate a cue, or to locate one ("If called by a panther, don't anther"). Second, it makes local environments heterogenous with respect to easily discriminated cues and the functional properties they signal. Deceptive fireflies mimicking the female signal to the male decrease the overall reliability of the signal/mate relationship, as the firefly environment becomes heterogenous with respect to the "species-specific" signal. Furthermore, they impose a cost on probing.

Let's pull all this together. Godfrey-Smith discusses the evolution of simple detection mechanisms, arguing that they evolve in environments which are heterogenous. Environments vary in ways relevant to the organism, and to which the organism can respond. Godfrey-Smith models the evolution of signal pick up, suggesting that it's a function of the reliability of the signal, the benefit of acting on it, and the cost of error. So he considers varied environments, but ones in which the variation is signalled to the organism. These environments are *transparently heterogenous*.[3]

I have used the idea of situated agency to stalk the shift from these simple systems to more complex ones in which organisms can "see through" variant or misleading proximal stimuli to the relevant features of their environment, through having the capacity to track particular features of their environment in more than one way. In *translucent worlds* there is a complex relationship between incoming stimuli that the organism can detect and the features it needs to know about. In these worlds there is selection for representation not just detection. In turn, I have suggested that generalist organisms will typically live in partially translucent worlds, and that hostility, too, has the effect of making critical aspects of an organism's environment translucent.

No-one would suggest that translucence is sufficient for representation. As the parasites and predators that crack the ant recognition and response system show, plenty of arthropods and other organisms are cue-bound with respect to their enemies. Nonetheless, if representation is within the range of evolutionary possibilities for the lineage, translucence will select for it. We need representation rather than detection in certain kinds of complex environment; those where the mapping between biological significance and proximate cue is complex. This complex of ideas is not easy to test, but I think it's a very plausible model of the evolution of belief-like states. It is easy to see how the capacity to represent key aspects of your environment might be a "fuel for success." One point of Godfrey-Smith's project is to synthesise the pragmatist idea that the point of cognition is the appropriate control of action with the reliabilist idea that the point of cognition is to accurately track the world. Accurate tracking is a means to appropriate behaviour, and helps explain the success of that behaviour. In turn, that pattern explains the evolution of the cognitive mechanisms that construct representations which track the environment.

III NIKE-ORGANISMS

In thinking about the evolution of cognitive capacities, Godfrey-Smith is typical in focusing on our capacity to represent our world; that is, on the evolution of belief. But what of desire? What explains the evolution of preferences? We digest, breathe, and vary our heartbeat rate without any cognitive representation of the metabolic needs these activities service. Over longer time periods, organisms partition their metabolic resources among growth, somatic maintenance and reproduction, quite often in somewhat flexible ways, again without any internal representation of the needs these trade-offs serve. The gonad weight of birds changes dramatically with the season, as does, in migrants, their body weight and composition as they build up their resources for migration. So it is not always necessary to represent your needs in order to act on your needs. Hence many organisms do not have preferences.

In McFarland's useful terminology, the "cost function" of an organism is its real trade-offs between dangers and resources (McFarland

1996). He points out that many organisms do not represent their own cost function. Instead of representing their needs they simply have a built-in motivational hierarchy. These are *Nike-organisms*; they just do it. For them, motivation derives directly from the value of internal variables that are keyed to internal metabolic conditions of the organism. Their behaviour is a response to a mix of external and internal signals. They act on hunger, thirst, and environmental states. They forage when their internal food reserves drop, and the value of a specific internal variable is cranked up. They drink on internal signals of dehydration. Kirsh calls this the "capacitor" model of motivation (Kirsh 1996) and it is perhaps most vividly illustrated by the "hydraulic" models of motivation in early ethology, and illustrated below (Lorenz 1950). Motivating impulses of various kinds ("action specific energies") are released by specific stimuli. The greater the energy build up, the lower the threshold on the releasing stimuli.

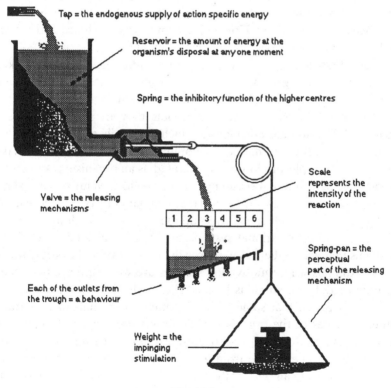

Fig. 11.1

The existence of Nike-organisms shows that desires are not mandatory. What then selects for the evolution of desire? Under what circumstances is it important for the organism to have a preference structure? For preferences have a cost. Any representation entails the possibility of misrepresentation, so representing the cost function brings with it the cost of misrepresenting it. Humans, very obviously, have embraced the opportunity to misrepresent their cost function many times. We cannot assume that there has been selection for the capacity to represent cost functions. Desire might be a side-effect. Perhaps representing motivational states of yourself is a side-effect of representing the motivational states of others: prey, predators, conspecifics. Perhaps it's a side-effect of a fancy capacity to represent the world. But if a preference structure is an adaptation, it seems most likely to be an adaptation to the complexity of *choices* facing the organism. That is, the descent of desire is a response to complexity, but complexity in the output side: complexity of responses to the environment. So once we shift focus from the evolution of belief-like states to the evolution of desire-like states, the environmental complexity hypothesis faces a different test. How is the complexity of choice and action related to the complexity of the environment?

Let's first consider the nature of the difference between Nike-organisms and organisms with preferences. "Hydraulic" models of motivation are just models; their implementation details are not meant to be taken seriously. But even as models, they are no longer widely accepted. But homeostatic models, which fit equally easily into Kirsh's general conception, remain quite popular. On these models a behaviour – for example, drinking – is generated as an organism responds to a variance in its internal economy between an ideal value of some state and its actual value. At its simplest, an animal detects, say, the partial dehydration of its cells, and drinks till the signal of dehydration ceases. Animals with motivational systems of this general kind differ from animals that act on their preferences in at least two and possibly three ways. Animals whose motivation systems are well-captured by homeostatic models are stimulus bound. Their behaviour is a response to stimuli. It is just that some of these stimuli are signals of important internal states of the organism while others are signals of important external states. So (first) one transition relevant to the evolution of preferences is the change from *internal detection* to *internal representation*. However, an internally represented motivating state is still not a preference, for preferences are typically representations of an animal's environment, but representations of how it might be, not of how it is or is taken to be. So (second) the evolution of preference from capacitor systems of motivation involves not just the emergence of representation, but also shifting its content outside the organism.

How and why might this happen? The internal environment of an organism is not hostile, so hostility does not drive organisms to represent rather than merely detect their metabolic states. But in discussing the shift from detection to representation, I noted that one problem with single channel tracking is feedback. If a signal changes slowly in response to changes at its source, it will not provide good feedback. Manning and Dawkins point out that cell dehydration has this feature. Though cell dehydration is important in initiating a rat's drinking, it is not a good mechanism for telling the rat when to stop drinking. For rehydration takes ten to fifteen minutes (Manning and Dawkins 1992, pp. 87–8). When an animal needs a halting criterion as well as an initiating criterion, *the mere cessation* of the initiating signal often will not do. We can see here a circumstance which might select for a change from internal detection to internal representation. Start signals often will not be good as stop signals and vice versa. The rat can control its drinking adequately by tracking an internal condition, though using two different cues. But for some behaviours, the halting criterion might be external not internal. To know when to stop, the animal needs to track *a change* in the environment. Anti-predator behaviour, I suggest, will often have its "off" switch tuned to an external environmental change.

So organisms with preferences contrast with organisms with capacitor-style motivations by representing rather than merely detecting internal states which produce their behaviour, and perhaps at the most primitive level of preferences by representing environmental changes as halting criteria in the control of behaviour.[4] Finally (third), we often think of preferences as forming a *system*, a preference order which enables an agent to choose amongst competing means and competing ends. On this view, organisms have preferences only if their motivations satisfy the conditions of commensurability, transitivity and so on that enable utility function to be calculated. Moreover, motivations can themselves be the object of motivation: we have preferences about our preferences. Since it is not obvious that humans fully satisfy these systemic aspects of preference structure, I will assume here that they are evolutionarily recent, and are inessential to the evolution of preferences from simpler motivation systems.

I have assumed that the complexity and control of behaviour must be central to the evolution of preferences. But behaviour can be complex in at least two ways. Behaviour can stand in a more or less complex relation to its initiating stimulus. A cock that gives false food calls only when a female is present, but when she is unable to see that the calls are misleading behaves more complexly than a cock that cannot tune its behaviour to such a subtle feature of its environment. More obviously, the behaviours themselves can vary in complexity. Thus Povinelli argues that arboreal locomotion increases significantly in complexity as organisms become heavier. In part, this increase in complexity is in the stimulus/behaviour relation. As animals become heavier, branches deform under the animals' weight, and so line-of sight routes cease to be possible. The relationship between what the animal sees and the route through which it moves becomes more complex. But actual movement itself becomes more complex: it is less stereotyped and involves multiple support points (Povinelli and Cant 1995). To a first approximation, complexity of the first kind – of stimulus/behaviour relations – is more tied to the evolution of belief-like representation, and complexity of the second type to preference-like representation. For the second kind of complexity poses problems of control.

We can draw the same distinction amongst the different ways behaviour can be complex by imagining two different ways an animal's behaviour can become stimulus independent. One is via belief: organisms become stimulus independent as their behaviour becomes increasingly affected not just by their motivating states and their registration of their current environment but by their belief-like states. Their stored information becomes increasingly relevant to their behaviour. But there is another potential route, running through the evolution of a preference structure. In the same circumstances different preferences will generate different actions.

I have suggested that motivations shift from internal detection to internal representation through the importance of feedback in controlling behaviour; in the simplest cases, when switching the behaviour off depends on a different cue than switching it on. Here the problem is to control a single element (drinking) in the rat's behavioural repertoire. But there is a second way in which control becomes complex, and one which may be even more important. Animals behave complexly in this second sense to the extent that we are unable to model their behaviour as consisting of a repertoire of discrete capacities, each with its own initiating, feedback and halting requirements. Motivating conditions can be relevant to *more than one element* in an animal's behavioural repertoire. As motivational states interact with one another and with other internal states, the integration of behaviour becomes more complex. As a consequence, the system as a whole will register the existence of the motivational state in its behavioural dispositions in more than one way. For internal states will impact on behaviour through more than one causal route. In virtue of complex control the same motivational state can (help) produce different behaviours. For its impact will be modified by other internal states. The relations between environmental input, motivational states and behaviour will become less direct. Contrast this situation of complex control with the simple homeostatic models. These are classic instances of capacitor models of motivation, and they are models in which the relation between motivational state and behaviour is simple and single-tracked. We shift beyond mere motivations as a given motivational state – say, a given value on the lust capacitor – can interact in different ways with other states to produce different behaviours. The state contributes to the shape of different behaviours rather than succeeding or failing to get its hands on the rudder. "Fight or flight" decisions may well be motivation without preference, for one motivation can simply be trumped and is silent in behaviour. But if the "flight" motivation affects the tactics of the fight without causing flight, we are shifting towards more complex control. Ethologists since Tinbergen and Lorenz have noted many behaviours apparently under complex control of this kind. The "tactical deception" literature provides many examples. One clear example is the suppression of copulation cries by chimps who rightly suspect intervention if their activity is discovered. The "mate" routine is not running independently of the animal's other motivation/action systems. Control has become more complex.

Hence I think more complex behaviour evolves at least in three ways: (i) increasing information about the world, including information not given in immediate perception, liberates behaviour from stimulus control, making behaviour less predictable given only information about a creature's immediate environment; (ii) the need for independent initiating and halting criteria, and more generally the demands of feedback, require animals to multi-track both their own internal states and changes in those states and their environment; and (iii) behaviour becomes more complex as metabolic motivators become relevant to more than one element of their behavioural repertoires. Under such conditions, control becomes less discrete, less modular. My hypothesis is that (ii) and (iii) are central to the evolution of desire-like states from brute-causal motivational states. In the rest of this chapter, I propose to consider (ii) in a little more detail.

I suggest that preference structures evolve as an animal's behaviour becomes more intricate and less stereotyped. A squirrel monkey scurrying along a branch may not need preferences; arguably, an orangutan does. For some animals need to represent instrumental goals, or subgoals, because they are faced with a variety of means to one of their ultimate ends. You need to represent your goals when you need to be able to represent subgoals and choose between them. Organisms that *plan* need to know what they want. The most developed case for planning in non-human animals has been constructed by Richard Byrne through his idea of a "behavioural program" (Byrne 1995, p. 68). He introduces this idea in discussing the vexed issue of primate imitative capacities. Though there is plenty of anecdotal evidence of great ape imitation, experimental evidence for imitation is surprisingly thin (Byrne 1997; Russon 1997). He points out that behavioural routines may have an overall functional organisation, rather than consisting of a mere chain of independent behavioural atoms. Gorilla food preparation illustrates this idea of a programme. Since they often eat thistles and other rather awkward plants, gorillas often need to do a good deal of manual processing of their food before they can eat it. This processing is quite complex, and involves a division of labour between the hands that changes through different stages of the processing.

Byrne's idea was that if skills depend on behavioural programmes, imitation can involve copying that programme rather than a specific motor pattern. If young gorillas acquire this ability through imitation, they may be copying the programme rather than the motor sequences. Indeed, if an animal has the representational capacity to represent another's behaviour as a programme, we might expect it to use this representation to direct its own attempts at the same behaviour. For direct parent/child behavioural copying will often not be a suitable vehicle for social learning. Adults differ from their young in size and strength. But they also differ in co-ordination, for many adult motor subroutines are already assembled and automatised. So it would often be very difficult for a juvenile to use adult motor routines as templates for their own behaviour even if they can manage the required transformation in point of view.[5] True imitation may be rare because adult motor templates are not useful, and few animals have the representational power to develop an abstract functional representation of a skill.

Byrne argues that the primate lineage has seen three major episodes of cognitive evolution. One distinguishes the haplorhine clade (apes and monkeys) from the strepsirhine clade (lemurs and lorises); a second distinguishes the great ape clade from the monkeys; and a third singles out the hominid lineage. He thinks that the first of these episodes is driven by social complexity, but he argues that there is little reason to think that the evolution of the distinctive capacities of great apes are so explained. Great ape social groups are themselves of very varied size and apparent complexity. Their social groups show no obvious overall increase in complexity over those typical of haplorhines.

So Byrne argues that the distinctive feature of great ape cognition is the capacity to plan. In his view, this is manifest in their complex processing of food (as in the gorilla example above, but it's typical of orangutans as well), in tool construction and use (chimps), in bed construction (most great apes), and in unstereotyped and complex motion through an arboreal environment (orangutans). The same seems to be true of the use by one animal of another as a social tool, as they form and exploit alliances. When the top-ranking chimp tolerates the number three having sex, because he needs support in a coalition against the number two, the first is preparing to use the third as a social tool, and that needs goal representation as well as world representation. The alpha chimp must represent what he himself wants to do. These activities all involve co-ordinated actions unfolding over time. They require an ape, "in other words, to *plan*" (Byrne 1997, p. 21).[6] Byrne goes on to suggest that the representational capacities required here are then redirected onto physical objects and other individuals. But we do not have to accept this extension of Byrne's idea to accept that flexible plan building requires the evolution of preference structures.

How does this view of the evolution of preference structure bear on the environmental complexity hypothesis and on the extent to which the evolution of beliefs and preferences are linked? On this picture, we certainly would not expect the complete decoupling of the evolution of belief-like structure from desire-like structure. If Byrne's picture of great ape cognitive evolution is right, environmental complexity is part of the story. Gorillas and orangutans have evolved behaviour programmes because of hostility: their food is defended by "spines, stings, hard casing, tiny clinging hooks" (Byrne 1997, p. 19). Tool manufacture and use, and perhaps social tool use too, strongly suggests that they categorise some aspects of their environment in functional terms rather than just in terms of sensory similarity. There is, for instance, quite wide physical variation in termite fishing twigs. But the complexity of the environment is not the whole story. The pressure for planning derives as well from apes' large body size and unspecialised digestive systems. These factors intensify the nutritional challenge facing the apes. These are internal features of the lineage. So too are features of great ape social life. The life history patterns of great apes may also be relevant: they have long periods as juveniles in which they have the opportunity to learn from adults.

So it seems unlikely that anything like preferences could evolve in the absence of belief-like structures, for environmental complexity and its representation does seem part of the story. Moreover, the stimulus-independent behaviour that the evolution of preferences would produce would in itself make the environment translucent for any animal that needs to predict what its fellows are likely to do. It is much less obvious that the converse is true: animals might have belief-like representational capacities without having anything much like preferences. If, for example, the social intelligence hypothesis is right about the haplorhine cognitive burst, they will need to represent their social world. For such monkeys form, maintain (often through grooming), and track long-term social relations, since gains and losses in social encounters often depend on these relations. As Byrne notes, in this clade triadic interaction is often critical. It does not follow that such monkeys need complex control mechanisms. Once you have successfully tracked your environment, perhaps there is no problem deciding what to do and how to do it. Hauser reports that rhesus monkeys that discover food caches scan their environment before either calling to advertise the food or silently eating. Those caught eating silently are likely to face attack, so Hauser suspects that prior to food consumption, discoverers are searching for both friends and foes[7] (Hauser 1997). Here, plausibly, we have a belief-mediated route to a looser link between stimulus (sighting food) and behaviour (food calling). If the rhesus monkey can determine what situation he is in, the decision problem is solved.

I have been discussing a specific hypothesis about the evolutionary origins of planning, and hence of preferences, in the great ape lineage. This might suggest that the evolution of preference-like representations is very rare, restricted to a single tiny twig in the mammalian clade. There are possibilities which suggest a broader distribution of the capacity. Change over time might select for the capacity to represent your needs by selecting for the capacity to act now to satisfy urges that you do not now have. Many animals store food rather than lose interest in food when satiated: leopards and some canids conceal unconsumed parts of kills. Squirrels and a good number of bird species cache food too. So, does future-oriented behaviour show representation of goals? Not by itself. But perhaps it does when it's flexible. Do acorn woodpeckers store acorns when they live in less seasonal environments with a steady supply of food all year around? Do leopards and dogs continue to cache food when food is freely and predictably available? If not, then it's likely that they represent their own needs. If the capacity to act for the future is flexible, and is sensitive to what the organism is really likely to need in the future, the case for the representation of goals looks strong. Environmental variability in the form of boom and bust food cycles generates the need to represent intermediate goals.[8]

This chapter is a first pass at the problem of the evolution of preferences. At best, it makes the problem explicit, and sketches a few hypotheses about that evolutionary process. But I have tentatively endorsed a version of the Environmental Complexity Hypothesis about the evolution of belief-like states, and I have argued that a certain species of complexity – informational translucence – is important in the shift from detection to representation. So how does preference – if my speculations are along the right track – fit into the picture? I do not think the answer is obvious. For we need to understand the relationship between behavioural planning and environmental complexity. Is planning necessary, or desirable, only in *certain kinds* of complex environments? At the very least, Byrne's examples make clear environmental complexity is one factor in demanding complex control. It may also be important in explaining the process through which behavioural repertoires become less modular, though that process remains murky. It is much less clear that environmental complexity is the whole story. The complexity of the environment seems to play no obvious role in the problem of knowing when to stop; when to stop drinking, when to stop burrowing and make the nest chamber; when to stop laying and start brooding, and so on. Some control problems, in other words, seem to be generated by features internal to the lineage.

So while the Environmental Complexity Hypothesis offers a plausible model of the evolution not just of simple behavioural plasticity but also of more complex genuinely representational states that are ancestral to belief, it is much less clear that it is the key to the evolution of preference structures.[9]

NOTES

1. How do I count channels between an organism and an environmental feature? Though I have no formal definition to hand, I should emphasise that channels are not sensory modalities. A baboon that monitors a rival by visually reacting to facial expression, body posture and relative location is tracking intent through three cues not one. See also Chapter 12.

2. Godfrey-Smith (personal communication) has urged upon me Dretske's idea that the multiple channel condition should be used to solve the distality problem rather than drawing the distinction between perception and detection (Dretkse 1988). Thus perceptual constancy mechanisms show that the representational content of perceptual states are features of the environment, not features of the proximal stimulus. My response to Godfrey-Smith here is programmatic. In a series of papers (including this one), I have tried to show the theoretical productivity of drawing the distinction this way; see also Chapter 8; Sterelny 2000a.

3. More exactly, Godfrey-Smith does not explicitly distinguish between transparent and translucent environments, but his examples are of transparent environments, for he sets himself to explain the most basic behavioural flexibility.

4. I would assume that tracking changes in the external world will be important in other aspects of feedback, too.

5. It is this transformation in point of view that has lead some to argue that true imitation requires metarepresentational capacities, and that is why it's so rarely found except amongst humans. I argue against this view in Chapter 10, while agreeing that imitation is cognitively sophisticated because it does require a functional rather than sensory representation of the skill imitated.

6. Prima facie tactical deception also involves planning. Yet Byrne does not regard it as confined to the great apes. So tactical deception outside the great apes looks somewhat anomalous on Byrne's view of the evolution of planning.

7. It is not clear how Hauser excludes the possibility of predator surveillance. He does note that solitary individuals who find food neither call nor are they attacked if discovered with food, though they are displaced. But he does not note whether they scan before eating.

8. At least for K-selected organisms. Obviously, there are many other evolutionary responses to environmental uncertainty of this kind. Many Australian animals are nomadic, or breed only in response to clear environmental signals of a good season, and this is presumably a response to the uncertainties of the Australian environment. Obviously features of the lineage itself are of

great significance in determining the character of the lineage's evolutionary response to uncertainty. But cognitive sophistication seems to be one possibility.

9. Thanks to Russell Brown, Fiona Cowie, and Peter Godfrey-Smith for their comments on an earlier draft of this chapter. I thank James Maclaurin for his reconstruction of Lorenz's hydraulic model.

12

The Evolution of Agency

You can't always get what you want
But if you try sometime
You just might find
That you get what you need

(Mick Jagger)

I WHY THINK?

Minds are control systems: they exist to drive behaviour. But some control systems are much more complex than others, and that is a fact that we would like to explain. One candidate explanation is the "Environmental Complexity Hypothesis": environmental complexity selects for behavioural plasticity (Godfrey-Smith 1996). According to this hypothesis, environments sometimes vary in ways relevant to the organism, and so responding differently in different circumstances can pay better than a fixed behavioural pattern. For an organism can optimise its behaviour for the specific features of its environment. To do so, the organism must have some means of tracking variation in the environment and matching its behaviour appropriately to that variation. The world must be complex in ways the organism can map, if minds are to pay their way.

Peter Godfrey-Smith explores this idea for very simple forms of plastic response; I am interested in extending it to more cognitively complex organisms. I shall discuss two ways of extending Godfrey-Smith's thesis, and I will explore the connections between them. In his examples, the epistemic problem facing the organism is relatively simple. I will consider the effect of more cryptic and confusing environments on cognitive evolution. I will also focus on the problem of control. In Godfrey-Smith's examples (and in many of mine), the organism faces only the task of selecting a simple response from a small menu of options. So I want to consider the evolution of more complex control: cases in which the menu is larger, and cases where the response itself is more multi-faceted. Like many others, I am interested in the evolution of intentional agency from simpler cognitive engines. But I shall focus not just on the evolution of belief-like representational systems, but also preference-like ones. For motivation can be grounded physiologically, in drives and sensations, rather than in desires. Perhaps sensations are representations of some sort. But desires are representations of a very special kind: they are motivationally salient representations of how the environment might be. Why would organisms evolve the capacity to form such representations?

The Environmental Complexity Hypothesis is an adaptationist hypothesis. It states that complex systems of behavioural control exist because they confer an adaptive advantage on the creatures with them. We do need an adaptationist explanation of cognition, for minds are expensive. Complex control systems require resources to build and house them. In general, adaptationist theories need to be developed carefully if they are to avoid declining into just-so stories, and this is particularly true of adaptationist theories of cognition. Cognitive complexity can be a bug as well as a feature. For behavioural flexibility risks error. No system of tracking, decision, and action can be error-proof. Thinking involves a cost of error: any organism that attempts (say) to discriminate between a harmless snake and its poisonous model – rather than treating all snakes as dangerous – risks producing its "that is harmless" response to the model. The more finely a behaviour is adapted to one state of the world, the less adaptive it will be if discrimination fails.

So if there is selection to tune behaviour to subtle features of the environment, it will be despite the cost of error. If the environmental complexity hypothesis is right, we must identify the environments which are complex, and complex in the right way, to make fine tuning worth that cost. The rest of this chapter unfolds as follows. First, I attempt to characterise a general feature of environments that selects for the capacity to track features of the world robustly; that is, to track them using a multiplicity of cues. I route this idea through the example of the social intelligence hypothesis. Here I develop further the ideas of Chapters 9 and 11. In doing so, I distinguish two aspects of cognitive response to environmental complexity. The first is tracking robustness. The second is response breadth – the sensitivity and variety of an animal's response to what it knows about. Broad response poses problems of control and motivation, for an animal with such capacities both has more options and is sensitive to more features of its world. Motivation is a central theme of this chapter, and the fact that it can be grounded physiologically, in drives and sensations, rather than in desires poses our problem. Since motivations can be based on a hierarchically organised drive structure in which some drives trump others, we face two questions. Why do preferences evolve and how can we tell, empirically, that they have?

Tony Dickinson has attacked this problem. He argues, perhaps surprisingly, that rats are intentional systems. They have both beliefs and preferences. In support of these ideas, he has developed an impressive body of experimental work. Nonetheless, I argue against his analysis. I doubt that rats are intentional agents. Having rejected Dickinson's line on preference, the paper concludes with a sketch of my own. In it I connect the two dimensions of cognitive response to complexity – robust tracking and broad response – to the problem of preference. I argue that a preference-based system of motivation is needed only for those animals with quite elaborate behavioural repertoires, and hence it's unlikely to be a common feature of how animals represent their world.

II TRACKING THE WORLD

Just about all organisms, by design, track some features of their environment. In their normal environment they respond adaptively with some regularity to that feature. However, in my view, an animal's capacity to manage its environment changes in very significant ways when it becomes able to track its environment via more than one kind of proximal stimulus (Chapters 9 and 11). An animal that tracks functionally relevant features of the environment in more than one way has more robust behavioural capacities. Arthropods are often efficiently cued into the relevant features of their environment, but they are often dependent on a single proximal cue. There is only one channel through which information flows from the world to the organism. Fireflies, for example, recognise potential mates by a species-specific pattern of flashes. But they have no other mechanisms that would enable them to recognise mates in different ways. Single-cue systems of this kind are vulnerable to exploitation. Thus one species of firefly is exploited by another predacious species that uses the species-specific signal as a lure (see Lloyd 1984). Capacities that depend on a single cue are *fragile*; those that depend on several are, I shall argue, *robust*.

This idea depends, of course, on our ways of identifying and counting channels between an organism and an environmental feature. There is no problem, of course, in counting cues when they rely on separate sense modalities: when vervets react to the voice rather than the sight of a leopard. But channels are not sensory modalities. A baboon that monitors a rival by visually reacting to each of facial expression, body posture and relative location is tracking intent through three cues not one. But two stimuli are never exactly the same. Two retinal projections of the silhouette of a hawk will always be different in some way. So even a cue-bound organism must and will generalise, treating physically different stimuli as functionally equivalent. So how can we distinguish stimulus generalisation from the use of several cues? Suppose we are interested in whether an animal detects predators by just a single visual cue. If the organism is cue-bound, we ought to be able to construct a single optimum cue, a visual stimulus the elicits the strongest response. As we deform the stimulus along various dimensions, the strength of the response should drop off. It might do so relatively smoothly in some respects. In others, there might be threshold effects – large changes in response over small changes in stimulus character. If we conceptualise this as a visual search space, with height in the space reserved for reaction strength, then for a single cued organism the space should contain just a single hump. Its capacity for stimulus generalisation is reflected by the shape of that hump: whether it has a large plateau on top; the dimensions in which the drop-off is smooth, and so forth.

This problem is empirically tractable. For example, there is a rich experimental tradition of investigating alarm calls, and the conditions under which animals call. So, for example, I am inclined to treat chickens' responses to aerial predators as cue bound. They respond to a hawk silhouette, but the apparent size and velocity of the silhouette also are important. Though shape, size, and speed all matter, chickens have a single peak in their search space (Evans, Evans, and Marler 1993). The search space of animals which are not cue-bound will have several humps. Moreover, with these animals, we should expect the relationship between stimulus and response to be less stable; the response strength should show more variability.[1] For that variation is the result of other cues adding to, or damping down, the response to threat mediated by any single cue. To the extent that response is dependent on these other cues, our first cue, by itself, will become a less good predictor of the organism's response.

Animals whose tracking capacities are cue-bound behave maladaptively if ecologically relevant features of their environment – their functional world – map in complex, one-many ways onto the signals they can detect. Such organisms live in *informational translucent environments*. If resources, dangers, allies and opportunities map in complex ways onto physical signals they can detect, cue-driven organisms' behaviour will often misfire. Such organisms face the problem that many different sensory registrations form a single functional category, and similar physical signals may derive from very distinct functional sources. In contrast, an environment is *transparent* if there is a single cue the organism can detect that reliably correlates with some need. Sexual receptiveness, for example, is often signalled transparently by females as they come into season.

Under what circumstances will different organisms find themselves in translucent environments? One reason is that members of the same species are often found in different environments. The greater the variety of environments in which an organism might find itself, the less it will be able to rely on constant physical signals to identify food, shelter and its other needs. When different populations of the one species occupy different niches, selection cannot predict the specific markers of the key environmental features of the organism's world. The relationship between the physical signals generalists can detect and the functional features about which they need to know is likely to be complex. Hence, the organism will need multiple routes to the environmental features of interest to it. So if organisms are environmental generalists, selection will act against cue-driven behaviour.

Some of the problems posed by ecological generalism might be solved just by the ability to recruit extra, supplementary signals of the feature of interest. If a rat can learn to associate a new smell and taste with the satisfactions of satiety, it can recruit those as new perceptual signals of food. However, hostility poses less tractable problems. These days anti-representational theories of mind are in vogue in the philosophy of psychology. Andy Clark has recently reviewed these ideas sympathetically, discussing models of adaptive behaviour not depending on internal representations of the external world (Clark 1997). But while these are interesting and suggestive, I think they are much more limited than Clark suggests. His central examples concern adaptive and flexible response to the contingencies of the *inanimate physical environment*. Clark discusses models of reaching for objects in the world; navigating around a cluttered environment; and of insect-like walking over uneven surfaces.

However, the epistemic problems facing real biological agents involve a far more difficult class of problems. The terrain, for the most part, does not care whether an insect can walk over it or not. Terrain is not selected to actively sabotage or hinder an insect's progress. As Clark points out, an animal's ability to act in its environment is often "scaffolded" by the structure of the environment itself. Thus those sceptical of representational theories of cognition often speak of "the world being its own best model" and of "storing information in the world." These metaphors depend on the environment being benign (as it will be if, like a termite mound, it has been adaptively shaped by previous behaviour) or, at worst, indifferent.

Cue-driven behaviour offers a plausible picture of interaction with an indifferent physical environment. But an animal's predators, prey, and competitors are under selection to sabotage its actions. They degrade the reliability of simple signals through concealment, camouflage and mimicry, and they can impose high costs on active information search. Adaptive behaviour targeted on the inanimate world (and biologically indifferent parts of the animate world) can rely for control on simple cues of the relevant environmental structure. The world, in those respects, is transparent: the animal can easily see through proximal stimulation to the opportunities its environment offers. But much animal behaviour takes place in a hostile world of predation and competition.

In summary, Godfrey-Smith argues that the sensitivity of organisms to signals from their environment should depend on the reliability of the signal, the benefit of acting on it, and the cost of error. He considers varied environments, but ones in which the variation is signalled to the organism. These environments are transparently heterogeneous. I have discussed the evolution of a more complex phenomenon: cases in which an animal can "see through" variant or misleading proximal stimuli to the relevant features of its environment, through having the capacity to track particular features of its environment in more than one way. In translucent worlds there is a complex relationship between incoming stimuli that the organism can detect and the features it needs to know about. There are many reasons why an animal's environment might be translucent. The physical environment may be variable over short periods in ways that generate no simple and reliable physical signal: weather contrasts with seasonal change in just this way. But hostility is of particular significance as a cause of translucence. Hostility has the effect of making critical aspects of an organism's environment translucent. Not every animal that would be advantaged by robust tracking mechanisms will develop them. They must be within the range of evolutionary possibilities for the lineage. But there will be selection for robust tracking where the mapping between biological significance and proximate cue is complex.

III AN EXAMPLE:
THE SOCIAL INTELLIGENCE HYPOTHESIS

There is no single social intelligence hypothesis. Versions of the hypothesis vary (i) on which primates the hypothesis applies to; (ii) the extent to which competitive interactions are emphasised at the expense of co-operative ones; and (iii) the extent to which the hypothesis is tied to the existence of a primate theory of mind (on this, see Chapter 9 and Sterelny 2000a). However, on every version of this view, a primate's life chances are critically affected by its ability to predict the behaviour of its social partners. In some circumstances, with some social animals, this presents no special problem. In co-operative interactions, when there's no conflict of evolutionary interest between the agents, selection might establish honest and unambiguous signals of intent. Mother/infant interactions, and the chimpanzee "play face" might exemplify honest signals and hence an environment which is transparent with respect to these aspects of social life. Honesty might evolve, too, as a result of Zahavi's "handicap principle" (Zahavi and Zahavi 1997): signals of health, vitality and strength might be too expensive to send for animals lacking these attributes. Only large strong stags can roar loud and long.

However, the thrust of the various social intelligence hypotheses is that for primates many problems of behavioural anticipation have no simple solution. The social environments of primates – or of particular primate species – are informationally translucent. For example, Tomasello argues that only primates adjust their behaviour to the relations between third parties. This awareness of third-party relations is shown in various ways: in the choice of allies; in the direction of aggression; in the disruption of potentially dangerous alliances; in reconciliation (Tomasello 2000). Now if one agent's behaviour is sensitive to these facts about his or her social environment, any attempt to anticipate that behaviour must, equally, be plugged into these features of the animal's social world. Considerations of this general type suggest primate behaviour usually cannot be read off any single simple behavioural cue. That has lead to the debate in the primatology literature on whether primates have a "theory of mind," and hence anticipate behaviour by understanding its psychological causes, or whether they are only "behaviour readers" (Heyes 1998).

This distinction is too crude, but the debate serves to illustrate the importance of two dimensions of cognitive complexity. The first of these is robustness: do primates track, say, the motivational states of others robustly? Or do they exploit only a single behavioural cue (in which case we could, indeed, regard them as behaviour readers)? The second is breadth: how nuanced is their response to the states they track?

Since many live in social groups in which others' behaviour is of critical importance, many primates must adapt their behaviour to the psychological states of other primates. Let's consider an example of such a problem. Franz de Waal discusses the problem of reconciliation after conflict for chimps (see de Waal 1982; de Waal 1989). Such reconciliation is apparently psychologically very important for chimps. But it is also very fraught, and the more serious the conflict, the more fraught reconciliation becomes. That is presumably in part because those seeking reconciliation are in a motivationally mixed state – the desire for peace mixed with hostile motivations – and a motivationally unstable state. But it is also fraught because in some circumstances, reconciliation and refusing reconciliation becomes a weapon of the ongoing conflict, with one chimp refusing overtures until the other clearly accepts inferior rank. So we can speculate with some plausibility that reconciliation may involve a "chicken" game: each of the rivals wants reconciliation, but they also want to grant it rather than seek it. Worst of all is to seek it and be refused. So when looking for the signs of approach, or at least willingness to be approached, in the other, it is important for the chimp to be right. And since motives are mixed, it is no surprise that although gestures seeking reconciliation are fairly standardized (an outstretched hand) there does not seem to be a single reliable signal. In de Waal (1982), in particular, de Waal charts a number of roundabout routes to temporary peace.

So let us suppose that chimps respond differentially to chimps that are motivated to seek peace. That is, they often recognise the clues, the signs that signal the motivation to reconcile. Their own behaviour is adapted to behaviours that are as a matter of fact caused by that distinctive psychological state. So suppose:

(i) one chimp always reads actions a, b, c, d, e . . . as actions of the same type;

(ii) actions a, b, c, d, e . . . are in fact always generated by a distinctive mental state Q; let's suppose Q is the desire for peace.

If chimps satisfy these conditions, they have at least the capacity to *track* the desire for peace. Their behaviour is adapted to that state. They recognise and respond to behaviours that are consequences of, and hence cues to, a particular psychological state. There is a flow of information about the motivational states of the potential peace-seeker to the mind of the responding primate. Suppose that a further condition is met about reconciling behaviour:

(iii) a, b, c, d, e . . . do not have any single simple sensory cue in common. There is recent work suggesting that chimps track visual attention by a simple cue, "face visible." Chimps, let's suppose, do not track the desire for reconciliation like this. Body posture, facial expression, vocalizations can all feed into a response distinctive of the desire to reconcile.

If the overtures that the chimp categorises together share no *single* distinctive sensory cue, the chimp is not stimulus bound with respect to this desire, but can track it via a variety of its manifestations. Generalising from this example, then:

a primate responds to the mental state of another if it can track some suite of behaviours that are actually caused by some specific mental state: for example anger or fear. They track fear, if they categorise together the spread of behaviours which are all as a matter of fact caused by fear.

If, as de Waal suggests, a bonobo appeases angry behaviour by trading sex for peace (de Waal 1989) it's *tracking anger*. We can investigate the robustness of this tracking both by testing for the variety of cues the bonobo uses, and probing their interaction. Perhaps different behavioural symptoms – posture, vocalisation, voice – act as independent sufficient conditions, analogous to a rat learning tastes of different foods. Perhaps they are not treated as individually sufficient, but each adds an independent and fairly invariant weight to the "Freddie is dangerous today" judgement. Perhaps the cues are not treated as having weight independently of one another: some will only be salient (or will be more salient) if others are present. Perhaps posture is discounted if there is a mismatch with voice or expression. As we shall see, there will need to be interactive effects between the cues if tracking is to be robust enough to unmask active and passive deception.

Animals' cognitive repertoires vary not just in the features of their environment they track, and the robustness of that tracking. The *breadth* of their response varies too, and this is my second dimension of cognitive complexity. Does the anger-reader adapt to angry behaviour differently in environments which cause that behaviour to be expressed differently? Does it respond differently, if the physical or social environment is different in important ways? Does it respond to anger differently, depending on its recognition of other mental states of the animal? Or is the "anger-behaviour" rule simple: run! This distinction defines two separate experimental investigations. One is robustness, to a first approximation investigated by the variety of observational cues it uses in tracking. The other is breadth: the extent to which the tracker's expectations about, and responses to, the agent's behaviour are appropriately modified by what else the tracker notices.[2]

We can then think of an animal's social intelligence developing via two sorts of behaviour rules. *Recognition rules* link a reader to tracked mental states. We search for the animal's recognition rules by fixing as far as we can the reader's environment, but varying behavioural cues of a single underlying cognitive state, to see whether the reader gives the same response to these different cues. *Output rules* govern responses to the states a reader can track. We probe an animal's output rules by fixing the reader's cue, varying the environment, and testing for different responses. I have formulated these rules through the domain of social intelligence, but the distinction is quite general. The classic experiments on chicken alarm calling were really *robustness* experiments. The experiments attempted to find a simple cue (a stylized hawk silhouette) that triggered alarm calling, with the intensity of the alarm dropping off as the silhouette shape was increasingly deformed (Evans, Evans et al. 1993; Evans, Macedonia et al. 1993). But much recent work on alarm calling are probes of *response breadth*. The idea is to test whether calling is affected by the existence, nature and situation of an audience. Does it make a difference if that audience includes a potential mate or a possible relative? Does it make a difference if the caller is safe before calling? If the audience is already safe? These are not probes of what the animal tracks and the cues they use in that tracking, but of what they can do with what they know. They are probes of response breadth. Vervets, for example, only call if there is an audience to warn, but they do not seem to take into account their audience's location.

This is a useful way of thinking not just about what animals know, but also about what they do not. While it is clear that social learning is very important in primate lives, it has been surprisingly hard to find convincing evidence of primate imitation. Thus, observation of a model informs chimps of the fact that a certain object offers them an opportunity: a box can be opened, and it has food in it. But they do not seem to learn from the model the *means* of exploiting this opportunity. It is a matter of continuing controversy whether this failure is real.[3] But if it is, is it a failure to track or a failure to exploit the information they have? Clearly, chimps are able to track perceptually the individual movements out of which a successful solution is built. But one possible source of failure is an inability to link in memory the individual perceptions into an overall representation of the model's performance. An alternative explanation might appeal to an inability to use that information – perhaps because it's coded in a modality-specific way – to drive behaviour.

It would not be easy to test between these options, but it would be possible. For example, we might video the model's performance and test to see whether the failed mimic can discriminate between the true model performance and various deformations and variations; both ones that are functionally neutral, and those that are not. If the chimp could (say) select the video of the successful performance from the larger class of actions, that would suggest that the chimp can remember the action sequence as a whole. If that is right, and the failure to imitate is real, then it is a consequence of an inability to use that information to shape its behaviour rather than lacking an ability to represent the action sequence.

Multiple tracking evolves, I am suggesting, as an adaptive response to a translucent environment. The very simplest form of multiple tracking, in which each signal is treated as a sufficient indicator of the feature of interest, might not help the organism solve the problems posed by environmental hostility. In a very minimal sense, an animal capable of forming a conditioned association is capable of multiple tracking a feature of its environment. Pavlov's notorious dog learned to track the arrival of food not just through the sight of it on its way but through the sound of a bell as well. But the dog, as its response shows, has no ability to use the flow of information down one channel at a time to check the reliability of the other at that same time.[4] So a stronger sense of multiple tracking involves some form of integration or cross-channel checking. Deception illustrates the special importance of cross-talk. For it can be unmasked by tracking motivation through several cues (Whiten 1996). Cheney and Seyfarth's vervets seem to have this capacity. Vervets use two acoustically different calls to announce the presence of rival, potentially hostile, groups of other vervets to their mates. Cheney and Seyfarth used a tape to generate false alarms by one vervet using one of these calls, and then tested to see whether the scepticism this generated extended to the other call type, which, indeed, it did. De Waal, too, has some nice, though anecdotal, evidence of chimps trying to suppress leakage – signs of anxiety in confrontations that are telltale cues that undercut an animal's advertising. No doubt this integrative capacity might be scaffolded by the previous evolution of an ability to add new signals to cue the animal to the presence of a critical resource. Adding extra triggers of a given action might thus pre-adapt the ability to use a number of independent, but cross-checked, information channels.

I have distinguished between two dimensions of cognitive complexity. One is focused on the input from the world: on the robustness or fragility with which an organism tracks key features of its environment. The other focuses on behaviour: of the breadth of an animal's response to that feature of its environment. Robustness and response breadth are sometimes tightly linked. Most obviously, they are connected through feedback. For some ways of tracking a feature of the environment are ill-suited for the control of action through feedback. For example, chemical cues have low transmission speeds, so it is hard to use change in such signals to modulate behaviour. So response breadth can be intrinsically limited by the animal's methods of tracking. Feedback, however, is an instance of a more general phenomenon: exploration. Parrots and many other birds that hunt wood-boring arthropods knock on branches, using sounds they make themselves to locate their prey. Endurance chasers like African wild dogs apparently probe herds of potential prey, trying to create behaviours that reveal vulnerability.

In general, animals do not wait passively for information to come to them. Sometimes the signals that animals use in robustly tracking a feature of their environment are a direct result of their own exploratory behaviour. Even so, there is no a priori reason for supposing that robustness and breadth are universally connected. A particular danger might be difficult to detect, but once detected the appropriate response might be obvious. Conversely, a female might find the readiness of a male to mate is signalled quite unambiguously. She may still have a hard time knowing what to do. She may need to balance considerations of mate quality with the availability of other mates and with resource issues.

In discussions of the evolution of cognition, the focus has typically been on the evolution of our capacity to represent the world; that is, on the evolution of belief and its precursors. But what of desire? What explains the evolution of preferences? In considering this issue, we will return to the connection between tracking and breadth.

IV   THE INTENTIONAL RAT?

We digest, breathe, and beat our heart without any representation of the metabolic needs these activities service. So it is not always necessary to *represent* your needs in order to *act* on your needs. With basic metabolic functions, an animal faces no real problem of selection. Respiration is not optional. But even when an animal must select from its behavioural repertoire it need not represent its needs. Motivation can be based on the strength of various internal drives. Bentham thought that human action was under the control of two "sovereign masters," the sensations of pain and pleasure. While this is too restrictive a view of human motivation, some of our action *is* motivated directly by sensation. Very likely, there are some animals all of whose actions have such motivations.

So why would more complex motivational mechanisms evolve, and what are their behavioural markers? McFarland has introduced some useful terminology to capture our problem here. In his language, the "cost function" of an organism is the real trade-off it faces between costs and resources (McFarland 1996). He points out that many organisms do not represent their own cost function. Instead of representing their needs they simply have a built-in motivational hierarchy. These are *Nike-organisms*; they just do it. Their internal metabolic climate controls the value of the internal variables which drive their behaviour, perhaps in conjunction with a few external cues. They act on hunger, thirst, lust, fear, and the like. They forage when their internal food reserves drop, and the value of a specific internal variable is cranked up. They drink on internal signals of dehydration.

So why have organisms evolved the ability to represent, and hence sometimes misrepresent, their cost functions? Tony Dickinson takes up the problem of preference in a series of papers with various co-workers. His guiding assumption, surely a reasonable one, is that the evolution of preference is tied to flexibility in behaviour. He argues that intentional agents can adjust their behaviour to changes in the value of resources in ways that do not depend on immediate sensation, and that intentional agents know about the causal connection between their acts and those acts' consequences. In his view, rats meet these conditions but simpler systems do not.

Nike-organisms' motivations are keyed directly to signals of physiological condition. Even so, their behaviour is not wholly inflexible. The same perceptual signals, combined with different drive states, will cause different behaviour. So will different perceptual signals combined with constant drive states. So Nike-organisms can respond adaptively both to variation in the environment itself and variation in their own physical needs. Moreover, they can learn by association. They learn that one response to an environmental signal is rewarded, whereas a different response to that signal is punished. So they show behavioural plasticity both at a time and over time. They are, as Dickinson puts it, *habit machines*.

Dickinson's project is to identify and explain the limits on the flexibility of habit machines. One of his most engaging examples outlines the "Rolly effect," Rolly being a Dickinson dog incapable of learning that he gets fed more speedily if he stays out of the kitchen (Dickinson and Balleine 1993, p. 280). The Rolly effect is not specific to Rolly himself: Animals in general turn out to be poor at learning not to approach food, as a way of getting food. As Dickinson sees it, the insensitivity of Rolly's behaviour to information about its effectiveness shows that his approach to food is not an intentional action; Rolly does not understand the connection between what he does and what happens, this failing one of Dickinson's tests of agency.[5] Despite his limitations in the kitchen, Rolly might aspire to intentional agency in other aspects of his life. But Dickinson goes on to describe a species of artificial agent, Norns, that are (in his view) in no way intentional agents. Norns have no instrumental knowledge of the outcome of their actions; they just perform whatever act has been "reinforced in the presence of the current stimulus input." They are homeostatic beasts: Reward is just drive reduction. As such, Dickinson argues that their behavioural repertoire is limited in an important way. Imagine a rat that gathers food in two ways: hunting for protein and gathering for carbohydrates. All its life, the rat has been hungry. It has never fed to satiety on either food. What would happen after the first time it is satiated on, say, carbohydrates. Will it hunt or will it gather? Real rats under these circumstances hunt: they solve *the forager's dilemma*. But Dickinson and Balleine argue that Norns, being habit machines, cannot fine tune their behaviour adaptively in this respect.

> During previous foraging episodes, both actions have been equally rein-forced by their appropriate food rewards in the presence of an input from the hunger drive produced by a deficit in both resources. As a con-sequence, the hunger-hunt and the hunger-gather connections have equal strength. Now, for the first time, the Norn experiences a novel state of carbohydrate satiety, a state that has no pretrained connections with either action with the result that the creature finds itself with very little inclination to perform either activity and, at best, vacillating between hunting and gathering. In the absence of knowledge of the causal rela-tionship between each action and the associated food, the Norn cannot choose to hunt rather than gather on the basis of the fact that protein should now have a higher goal value than further carbohydrate intake. (Dickinson and Balleine 2000, p. 5)

So Norns, but not intentional agents, are crippled in novel situations. I agree that it is important for us to distinguish empirically between habit machines and intentional agents. So Dickinson is scratching where it itches. But I do not accept his characterisation of the problem. First and most critically, Dickinson assumes a preference/belief linkage. His taxonomy of control systems recognises only three categories: com-pletely inflexible systems; habit machines driven by association, and intentional systems. His central contrast is between habit machines and intentional agents. Intentional agents have instrumental knowledge of the action/outcome relation: they understand the causal consequences of their own actions. And they represent the goals of their own actions. But this way of setting up the issue prejudges an important question: Can the evolution of beliefs be decoupled from that of preferences? Dickinson poses a methodological question: What behavioural abilities distinguish habit machines from intentional agents? And he poses an evolutionary question: How did intentional agents evolve out of habit machines? In both cases, he presupposes that there is no belief without preference and vice versa.

Second, the example on which Dickinson focuses does not seem to me to be diagnostic of the distinction he draws. For he begins his discussion of the evolution of preference with an example of aversion: his own acquisition of watermelon aversion. This is no mere anecdote, for his key experimental manipulations probe the consequences of induced aversion in rats. Aversion is a very striking phenomenon, and powerfully illustrates the connection between preference and affect that is at the core of Dickinson's view of preference. For, as he shows, the physiological basis of an aversion only becomes motivationally salient through affect, and that affect has to be induced through the agent having contact with the object of aversion. Thus Dickinson did not realise that he was watermelon averse until he re-experienced watermelons, and rats made averse to glucose water do not seem to realise that they are sugar-water averse unless they are allowed to have contact with sugar water. The affect induced by aversion is unpleasant, but the importance of contact in making a physiological change motivationally effective seems more general. Dickinson and his co-workers showed that rats need to learn through the experience of eating when hungry that eating a food while being food deprived is a particularly satisfying experience.

Yet though affect plays a key role in *establishing* a motivation, Dickinson argues that the motivation itself is an abstract representation of the value of the commodity, not the aversive experience which sets that valuation. Hence rats are motivated by preferences, not sensations. For if the physiological mechanism that underlies the aversive experience is suppressed *after* the evaluation has been made ("watermelon, yuck") the substance to which the rat (or the psychologist) has been made averse continues to be rejected. Thus Dickinson gave nausea-suppressing drugs to the aversive rat, and yet they still continued to reject sugar-water. So it is not the *experience of nausea itself* that motivates, but the utility function induced by that experience.

Thus Dickinson's case for the existence of a rat preference order rests on the capacity of the rat to adjust its behaviour to changes in the value of resources, and the decoupling of that change in behaviour from sensation. So on his view there are two critical differences between habit machines and intentional agents. Intentional agents have a utility function, not just drives. And they know about the causal connection between their acts and their consequences. His supplementary hypothesis is that affective experience plays a critical and indispensable role in this process of re-evaluation.

The connection between affect and goal re-evaluation is certainly striking. I would never have guessed that rats need to experience the satisfaction of eating when hungry for the value of food, when hungry, to be motivationally boosted. But nonetheless there is something strange about the line of thought here. It is very odd to take aversion learning to be a central exemplar of preference change. For, in humans at least, a most striking fact about aversion (and its presumed positive twin, which we might call "addiction learning") is that it is *subintentional*. Dickinson's aversion to watermelon is impervious to the knowledge that it was too much wine, rather than too much watermelon, that caused his illness. Moreover, unlike most of our desires, watermelon-avoidance is not open to being *re-re-evaluated*. It is cognitively impenetrable (Balleine and Dickinson 1998; Dickinson 1985; Dickinson and Balleine 1993; Dickinson and Balleine 2000; Dickinson and Shanks 1995).

What is true of Dickinson seems to be true of his rats. Aversions seem to me to be instances of motivation without preference. In an insightful discussion of the evolution of motivation, Sober and Wilson point to three factors involved in the recruitment of a motivator (Sober and Wilson 1998). First the evolutionary history of a lineage must make the potential motivator available. Bat ultrasound might be a very reliable signal of the need to take cover, and some species of moth have recruited this signal as a sign of danger. But phylogenetic constraints probably make this signal unavailable to many prey species. A second factor concerns reliability. Pain is a very *reliable* motivator. Pain, in particular acute pain, has few false positives: if you feel serious pain in, say, your knee, very likely there is damage of some kind. Equally, if there is significant damage, there will probably be pain. So false negatives are also rare. Moreover, there is a very stable connection between the specific feature pain detects, namely damage, and its fitness implications. Avoiding or reducing damage is nearly always a good idea. So the sensation of pain is a reliable signal of something of significance to the animal. Moreover, the behaviour pain motivates is normally appropriate to the state it signals. It tends to induce rest, attempts to protect injured tissues, withdrawing from injurious stimuli and the like. So it is not surprising that the sensation of pain remains an important motivator for us. No doubt a significant range of human activity is motivated by preference with respect to pain. But much is motivated by pain itself. In humans (and doubtless many other species of animals) pain and other sensations have not been displaced from motivational significance by preferences, including preferences about sensations.

In his response to these ideas, this was the critical point of Dickinson's disagreement. In his view, pain and other affective states motivate *only* via their effect on preferences. But affective states (however understood) are clearly not identical to preferences, since one can be motivated without affect and vice versa. Furthermore, preferences seem to be representationally rich and structured: one cannot want to keep well clear of that snake without the concept of snakes. So this view seems to commit Dickinson either to the view that only representationally sophisticated organisms have motivationally salient affective states, or to the view that preferences can be unstructured, non-conceptual representations of some kind. Furthermore, given that on Dickinson's own view affect and preference depend on distinct psychological mechanisms, he faces an evolutionary puzzle. If the mechanisms that generate affect evolved before those on which preferences depend, on his own account affect would be epiphenomenal. But on Dickinson's view, since preferences are grounded on affect, preference generating mechanisms could not evolve prior to those of affect. So how did evolution arrange for the simultaneous evolution of these distinct mechanisms?

These questions about reliability are closely linked to cost. Costs can be metabolic: the price of extra cortex, or new sensory apparatus. But there is also a cost of error. As Mick Jagger reminds us, we do not always want what we need. Human utility functions are often out of step with their cost function, sometimes with catastrophic results. Error can creep in three ways. There can be false signals, and a failure to signal: pain without damage; damage without pain. The signal can generate an inappropriate response: itches can motivate damaging scratching. The signal can have the wrong motivational strength: sometimes over-riding more pressing matters; sometimes failing to over-ride when it should. No doubt pain sensations sometimes cause all three errors. For example, pain might sometimes induce us to rest a muscle which should be exercised. But there is no reason to suppose acting on preferences about pain would be more reliable than action on pain itself. There seems no reason to believe that an abstract evaluation of pain has been recruited as a motivator. Indeed, suppression of pain sensation suppresses much of our behavioural response.

How might these considerations about pain translate into rats' recalibrating their food preferences? A preference structure may not be among the evolutionary possibilities open to the rat lineage. But if it were, would preferences evolve? I doubt it. First, it is not hard to conceive of a directly affective proximal mechanism that would explain the aversion phenomena. If taste sensations are at all modifiable by experience, then there seems no reason to suspect the rat of intentionality. Dickinson's experimental set-up suggests that sugar-water aversion does not operate by making rats feel nauseous in confrontation with sugar water.[6] But it may be that the aversive experience makes sugar water just *taste bad* (as he himself notes at one point).

Second, there is no reason to suppose that such a sensation-based mechanism would be more error-prone than an intentional one. Would the evolution of a preference structure bring a rat's action more reliably in tune with its needs: will it want what it needs and need what it wants? Dickinson suggests, in regard to the Norns, that motivational mechanisms based solely on drives – non-cognitive motivational mechanisms – doom an animal to very limited capacities to adjust to new experiences. They cannot solve the forager's dilemma. Hence the Norn architecture would generate false negatives: They would lack motivation to forage for protein, even though they need it. They would fail to want what they need. Dickinson's suggestion seems to turn on an impoverished menu of drives and their associated sensations (on this, see Spier and McFarland 1998). Norns are helpless in the novel circumstance of satiation because they have come equipped with a single hunger drive. This drive is extinguished by eating either protein or carbohydrates, and that is why the Norns fail to switch adaptively to protein search. But suppose that there are *deficit-specific hungers*. Carbohydrate-need drives carbohydrate-hunger; protein-need drives protein-hunger. Gathering has tended to extinguish carbohydrate-hunger, and hence there are associative connections between carbohydrate-hunger and gathering; hunting has tended to extinguish protein-hunger and hence there are associative connections between protein-hunger and hunting. Through its history, the rat has pretty well always been both carbohydrate and protein deprived. So both hunting and gathering have been reinforced, pretty well as much as one another. So instead of a utility function, false negatives can be avoided through a richer menu of drives.[7] I think, then, that rats are subintentional systems; they do not labour under the mixed blessing of a mentally represented utility function.

V YOU CAN'T ALWAYS WANT WHAT YOU NEED

The status of drives and sensations in philosophy of psychology is controversial. Drives, I take it, are typically conceived of as a non-representational account of motivation. That is clearly true of the hydraulic models of motivation in early ethology, and seems equally true of the homeostatic models that replaced them. In this context, motivation is more often discussed in terms of arousal and of the role of the endocrine system rather than in information-processing terms. So insofar as motivation can be understood in terms of drives, motivation is physiological rather than representational. The status of sensation is more ambivalent. Many philosophers of mind have assumed sensations are not representations. They are simple intrinsic properties of some kind. But others have offered representational accounts of sensation. For example, Paul Churchland has argued for a vector-coding account of taste sensations. On this issue I can afford to be neutral. For preferences, I take it, are not just representations but representations of a particular type.

If folk psychology is right, humans are intentional agents. We are creatures with both beliefs and preferences. There is, of course, a serious dispute about what this amounts to. But for the purposes of this chapter, I take no stand on the Fodor/Dennett dispute on folk psychology. Dennett sees folk psychology as giving the "performance specifications" of intelligent behaviour. Folk psychology specifies the capacities of an organism that can intelligently act on the information available to it on the basis of its needs. Fodor does not disagree with Dennett's assessment of the abilities of intentional systems but he sees folk psychology as first and foremost a hypothesis about the cognitive engineering that explains those capacities. We have a language of thought, and beliefs and desires are sentences in that language of thought. Dennett, of course, accepts that some representational and computational capacities underlie and explain the information-using competence of intentional systems, but he regards folk psychology as ex officio neutral on their specific nature. In what follows I go beyond this neutrality only in assuming that paradigm intentional systems have some representations that are not functionally tied to specific acts, and others that do have a specific-action generating role.

Not every representational system is an intentional system. Ruth Millikan notes that when organisms are equipped only with relatively simple means of representing their world, we cannot draw a distinction between representations that merely report how the world is, and representations that direct behaviour (Millikan 1989). Cockroaches escape from predatory toads by detecting their presence from the wind gust caused by the movement towards them of the striking toad's head. They are equipped with antennae ("cerci") covered with hair-like wind detectors. When these register a wind gust of appropriate speed and acceleration, the cockroach turns away from the direction of the gust and scuttles for safety in that all too familiar way (Camhi 1984, pp. 79–86). Since there is such a tight linkage between what is detected in the world, and what is done, it seems arbitrary to translate the cockroachese thought as "Shit! Toad ahead!" rather than "Break left and run!" or vice versa. Amongst our other contrasts with cockroaches, within our mental representations there is a distinction between reports and instructions.

Beliefs are a "fuel for success": they are an information store about the world that advantages the animal in many different actions, but they are not tied to specific behaviours (Godfrey-Smith 1996). Preference, on the other hand is tied to specific action. As Millikan points out, the function of the desire for a good feed of steak is to bring it about that you are so fed. So the evolution of intentional agency involves the formation of world representations which are functionally decoupled from any specific action, while being potentially relevant to many. And it makes motivation cognitive. It brings motivation under the control of representations of the external world. As Fodor has repeatedly argued, the representation-forming mechanisms that support human belief and preference formation are systematic. They enable us to represent a wide, perhaps unbounded, array of situations both in a motivation-neutral way (in belief) and in a motivation-driving way (in desire). So the evolution of a representational system for encoding preferences makes the potential motivational base of human action extraordinarily wide. We can want, and act on wanting, an extraordinarily wide range of states of affairs. In doing so, our actions are no longer under immediate affective control.

I argue in Chapter 11 that preference involves the externalisation of motivation: The animal tracks not just how the world is, but also the changes the animal can affect in its world. As I have noted, Ruth Millikan identifies the representational content of a preference with the change in the world that preference brings about, if the animal acts on it in biologically normal conditions. There are problems with an attempt to define the representational content of the full range of human preferences by direct appeal to their biological function. Many human preferences are novel and some would require physically impossible acts. In these cases the idea of a "biologically normal condition" may be undefined (Sterelny 1990a). But Millikan has surely identified the core biological function of preferences. Her account may not apply to the full range of human preferences. But it would apply to the rat, if rats have preferences. So the evolution of a preference structure sensitises the organism to new features of its environment: the changes the agent itself causes in its environment. Moreover, the evolution of preference liberates motivation from its dependence on immediate affect, and hence frees the animal from phylogenetic constraints on its menu of distinct sensations and drives. An animal can learn to associate internal rewards with new stimuli. A rat can learn that sugar water now tastes horrible, or that a McDonald's hamburger bun tastes like food. But new drives and new sensation spaces can only be assembled over evolutionary time. If the cognitive mechanisms which assemble preferences are at all systematic, an animal so equipped can build new motivators in ontogenetic time.

In my view, the evolution of belief-like representation is not necessarily linked to the evolution of preference-like representation. An organism might be able to track features of the world in an action-independent way while its motivation system relies on direct signals of metabolic condition; drives and sensations. The reverse is not true: decision problems only become difficult when the animal can discriminate between many different situations. For only then does the problem "What do I do now?" have the potential to arise. So an ability to track many different aspects of the environment is a necessary, though not sufficient, precondition of the evolution of preference. Response breadth is an important precondition, too. Broad response capacities scaffold the evolution of preference. An animal only needs to control its actions through representations of the external world when it faces many choices. A baboon that acts in only one way when it detects hostility from a superior has no decision problem. A baboon that takes into account the number and location of his friends and those of his rival; the physical geography of the interaction; the value of a resource;

and perhaps other factors must clearly have a larger behavioural repertoire in response to his detection of a hostile superior. It might not yet
be very large. These other aspects of the situation might just feed into
a binary decision: concede or resist. The source of immediate motivation might still be non-cognitive: a competition between greed and fear.
But if cognitive evolution has given the baboon a finer discrimination
gradient – if it now recognises that there are several different kinds of
situation in which it faces a hostile superior – then the possibility is
born of the baboon developing an optimum behaviour for *each* of those
situations. As the range of potential behaviours increases, so the mechanisms of control must change. If the only actions a baboon contem-
plates in situations of resource conflict are to run or to grab, then the
decision might be made through the relative strength of two internal
drives. Such a picture of motivation looks increasingly less plausible as
the range of options increases. So even when broad response capacities are shown by the range of factors that feeds into the animal's
choice between a small range of options, that discriminative richness
has the potential to scaffold the evolution of a larger array of options,
which in turn require new motivational mechanisms.

Which animals need such mechanisms? Rats do not seem to need
to externalise their motivation. The rat needs to register the fact that
certain types of action are rewarded, and if they are to solve the
forager's dilemma, they need to keep track of the particular reward
a given type of act produces. In natural environments, these rewards
would be contingent on certain features of the environment. So the
rat might need to register the fact that a particular act is rewarded
only in certain circumstances. But what the rat does not need to
know – and what in fact the experimental set-up prevents the rat from
finding out – is *how* an act brings about a particular reward. The rat
does not need to keep track of how its acts change the world in a
rat-rewarding way. It just needs to know *that* the world has changed
in a rat-rewarding way. Yet preferences are motivationally salient representations of such changes. The tasks the rats are confronted with in
Dickinson's experimental set-up require them to track some subtle features of their environment, but they never require complex or subtle
responses to those features of their environment. His rats do not need
to show broad response to causal connection. Their world might be
complex and so knowing when intervention will be effective is a
difficult problem. But the interventions are few in number, simple
and structureless.

I doubt that a preference structure is needed to support interventions of that kind. My guess is that preference is linked to the complexity of choices which face an animal, and their relation to affective reward. I think there are three elements, at least, that make up the complexity of an animal's response to its world. One is the independence of behaviour from immediate affective reward and punishment. A second is the sheer variety of choices an animal faces: the effective size of its behavioural menu. A third is the complexity of control of specific behaviours; the extent to which the animal has to be aware of its own effects in the world. An animal might have a large repertoire of relatively simple interventions from which it must select.

Let me sketch this third issue of control a little more fully. If we need to explain the behaviour of a foraging rat as it trades off risk against the richness of a food source, we do not need to invoke preference to explain its increasing tolerance of risk as it grows hungrier. The sensation of hunger, we can suppose, is growing sharper and more insistent, tending to over-ride feelings of anxiety. However, we cannot appeal to a non-representational, merely affective account of motivation if the suppressed impulse adaptively affects the manner or mode of the unsuppressed behaviour. One clear example is the suppression of copulation cries by chimps who rightly suspect intervention if their activity is discovered. The "mate" routine is not running independently of the animal's other motivation/action systems. Similarly, it may turn out that there are examples of risk-sensitive foraging where risk affects not the rate of foraging but its manner: a monkey shifts to foraging, say, on the canopy of a tree rather than on the trunk. For such animals, control cannot be vested in just a hierarchical order of drives, one of which turns out to be the strongest. Moreover, it is hard to show an animal with a large range of options – one facing decisions about what to do, when, where and with whom – could act just by responding to drives or sensations. It would need both a large menu of sensations, and some way of balancing between competing urges. If an animal has evolved not just the externalisation of motivation, but also a way of ranking these motivations, it has evolved a preference structure. A utility function just is a translation of many different motivations into a common currency. That common currency is only needed, presumably, when there are many different motivations.

Another route to the evolution of preference is through the separation of action from affective reward. So seen, a preference structure is a very middle-class invention: it enables an animal *to delay gratification.* This is important in planning and in any other behavioural sequence extended over time, where only the final, successful, completion of the sequence brings rewards. I argued along these lines in Chapter 11, floating the idea that the most obvious need for preference-based motivation arises for animals which must plan behaviour. I take plans to be action sequences whose intermediate steps require detailed feedback. So understood, plans require, first, behaviours that are decoupled from immediate reward, and, second, require the animal to monitor not just the action but the change in the world the act causes. Such animals must somehow check that an internal representation of a goal is matched against the results of their action.

If the evolution of plans and preferences are tightly linked, preference may be restricted to the great ape clade. For Byrne argues that the distinctive feature of great ape cognition is the capacity to plan. In his view, this is manifest in their complex processing of food (as in the gorilla example above, but it's typical of orangutans as well), in tool construction and use (chimps), in bed construction (most great apes), and in unstereotyped and complex motion through an arboreal environment (orangutans). To treat planning as distinctive of great ape cognition we would, of course, need to have some reason to treat other apparently complex action sequences differently. Nest construction by birds, for example, involves complex constructions in which it is surely necessary for the builder to register the effect of his/her own work as the construction proceeds. There might be a case for treating nest building as non-intentional, perhaps because it is modular and automatic. The nests of a given species do tend to be very similar, hence the possibility of field guides to nests, and perhaps the whole sequence is "chunked" motivationally. But there is clearly a danger of primate special pleading here. There may, however, be other types of action where action typically brings no immediate reward. Animals cache food (though most such behaviour seems pretty inflexible); others make tools. More generally, many animals engage in epistemic behaviour, exploring their environment.

Putting all this together, motivations based on preference are distinct from those based directly on affective mechanisms in five important ways:

(i) While no doubt affect plays an indispensable role in establishing a preference structure, motivation is liberated from immediate affective reward. The point of an act is to change the animal's environment in the way specified by the content of the desire on which the agent acts. While that may involve an affective reward of some distinctive kind, it need not.

(ii) The evolution of preference sensitises the animal to a feature of its environment; the change the animal itself would make to its environment in satisfying its preference.

(iii) On the assumption that preferences are constructed from semantically simpler concepts, an animal that has preferences at all can have many different preferences: the range of preferences it might have – that lie within its representational competence – is large.

(iv) Hence, preference-based motivation liberates an animal from a few ontogenetically fixed sources of motivation. Such an animal can learn what it wants, not just how to get what it wants.

(v) If preferences are ranked, the animal has acquired a mechanism through which its behaviour can be made responsive to many sources of motivation, not just a few competing drives or sensory spaces.

The task will be to both develop empirical tests for the presence of these motivational systems, and a theory of the environmental complexity that requires them. I have identified a type of environmental complexity, translucence, which plays a special role in the evolution of robust tracking. Translucence, in turn, is largely explained by the effect of hostility. I would like to develop a similarly coarse-grained theory of the environments that advantage broad-banded response capacities and the externalisation of motivation, together with some characterisation of the lineages which have the capacity for such an evolutionary response. Of course, there is no guarantee that such a theory is to be had. For one thing, even if I am right in thinking that the cost of cognition ensures that complex cognition as a whole is an adaptation, various aspects of cognitive complexity might be side-effects. But there are other ways this project might go wrong. Even if broad-banded response, and externalised, affectively neutral motivation are adaptations, they may be adaptations for which there is no unitary explanation. Instead, these aspects of cognitive life might have evolved separately in different lineages through the interaction of features of a lineage itself and specific features of its environment. Byrne, for example, thinks that the great ape's ability to plan is an adaptation, but it evolved because they are both large and lack metabolic specialisations for diet. Their food requirements were large, and they lacked easy access to high quality food. So the selective forces which drove the evolution of the ability to plan depended on both specific features of the lineage itself, and specific features of the environment. It could turn out that the evolution of intentional agency is a medley of specific stories, with nothing important in common. But if we do not look for a more unified explanation, we certainly will not find one.[8]

NOTES

1. There will be some variability, of course, even for a cue-bound organism, for response strength will be affected by such internal factors as arousal.
2. Peter Godfrey-Smith has pointed out to me that we can collapse these into a single dimension. Instead of seeing a primate as having a behavioural repertoire involving choice from a variety of possible responses to a given situation (depending on circumstances), we can instead see the primate as circumstance-bound. In each situation that the primate recognises, he or she has only a single response. Cognitive complexity enters not as response breadth but as a discrimination gradient; many subtly different circumstances are recognised as different. This is an interesting challenge, though it's unclear to me that it effects the substance rather than the formulation of the line of argument here. In response (a) I conjecture that the item that I regard as the foreground – here noticing the angry chimp – is developmentally, evolutionarily, and motivationally prior to the background circumstances that may modulate the response. Thus the reason to act at all is the angry chimp, not the potential allies loitering under the trees. Take away the allies, and action of some sort is still needed; take away the chimp, and nothing at all need happen; and (b) nothing in Godfrey-Smith's challenge is specific to non-human animals. We could apply it to humans, and conclude that we too are circumstance-bound. Once our assessment of the circumstances we are in is fixed, so too is our action. That seems an implausible view of human action. So if it can be blocked in our case, perhaps the blocking manoeuver can be co-opted for primates in general.
3. For a recent review, see Boesch and Tomasello 1998. There are hints that ant fishing might be transmitted by imitation, since the technique used is constant within a group, but variable across them. Since both groups fish for soldier ants, there seems no obvious ecological explanation for the difference.
4. Moreover, an animal that can add cues only by association with a single original cue is ontogenetically cue-bound: all its evidence must be funneled through that original cue. It cannot learn except through that cue.
5. It is actually not obvious to me that Rolly's problem stems from a representational failure. Rather, it may be an "executive planning deficit." Rolly lacks impulse control. These experiments are very like those at which autistic children fail, when they cannot point away from a chocolate to gain it. Of course, a failure in representational ability would not be very surprising in Rolly's

circumstances. He only needs the capacity to represent a causal connection between act and outcome when that relationship is contingent; when it varies across different times and places. Approaching food normally is a *very reliable* means of getting to food.

6. Or, at least, such feelings of nausea are not necessary for the aversion to block consumption.

7. Dickinson is sceptical of all solutions using this strategy. In his view they have often been tried and they result in a baroque elaboration of internal drives. He argues that the root problem is that if we suppose that there is (say) a specific protein-deficit drive that is reduced only by protein intake, we lose our explanation of too many other learning phenomena.

8. Thanks to Peter Godfrey-Smith and Tony Dickinson for their comments on the ideas in this chapter.

References

Agar, N. (1993). "What Do Frogs Really Believe?" *Australasian Journal of Philosophy* **71**: 1–12.

Alexander, R. (1987). *The Biology of Moral Systems*. New York, de Gruyter.

Allen, C. (1992). "Mental Content and Evolutionary Explanation," *Biology and Philosophy* **7**: 1–13.

Allen, C. and Hauser, M. (1992). "Communication & Cognition: Is Information the Connection?" in Hull, D., Forbes, M., and Okruhlik, K. (Eds.), *Proceedings of the Philosophy of Science Association*. Vol. 2. East Lansing, Michigan.

Allmon, W. (1992). "Causal Analysis of Stages in Allopatric Speciation," in Futuyma, D. and Antonovics, J. (Eds.), *Oxford Surveys in Evolutionary Biology* Vol. 8. Oxford, Oxford University Press.

Astington, J. and Gopnik, A. (1991). "Theoretical Explanations of the Children's Understanding of the Mind," *British Journal of Developmental Psychology* **9**: 7–31.

Balleine, B. and Dickinson, A. (1998). "Consciousness: The Interface Between Affect and Cognition," in Cornwell, J. (Ed.), *Consciousness and Human Identity*. Oxford, Oxford University Press.

Barkow, J., Cosmides, L., and Tooby, J. (Eds.). (1992). *The Adapted Mind: Evolutionary Psychology and the Generation of Culture*. Oxford, Oxford University Press.

Bateson, P. (1976). "Specificity and the Origins of Behavior," *Advances in the Study of Behavior* **6**: 1–20.

Bateson, P. (1978). "Review of *The Selfish Gene*," *Animal Behaviour* **26**: 316–18.

Bateson, P. (1983). "Genes, Environment and the Development of Behaviour," in Slater, P. and Halliday, T. (Eds.), *Animal Behaviour: Genes, Development and Learning*. Oxford, Blackwell.

Bateson, P. (1991). "Are There Principles of Behavioural Development?" in Bateson, P. (Ed.), *The Development and Integration of Behaviour*. Cambridge, Cambridge University Press.

Bekoff, M. and Allen, C. (1992). "Intentional Icons: Towards an Evolutionary Cognitive Ethology," *Ethology* **91**: 1–16.

Bennett, J. (1991). "How is Cognitive Ethology Possible?" in Ristau, C. (Ed.), *Cognitive Ethology*. Hillsdale, LEA Press.

References

Bennett, K. D. (1997). *Evolution and Ecology: The Pace of Life*. Cambridge, Cambridge University Press.

Bickerton, D. (1990). *Language and Species*. Chicago, Chicago University Press.

Boesch, C. (1993). "Aspects of Transmission of Tool-Use in Wild Chimpanzees," in Gibson, K. R. and Ingold, T. (Eds.), *Tools, Language and Cognition in Human Evolution*. Cambridge, Cambridge University Press.

Brandon, R. (1984). "The Levels of Selection," in Brandon, R. and Burian, R. (Eds.), *Genes, Organisms and Populations*. Cambridge, Cambridge University Press.

Brandon, R. (1990). *Adaptation and Environment*. Princeton, Princeton University Press.

Brett, C., Ivany, L., et al. (1996). "Coordinated Stasis: An Overview," *Palaeo: Palaeogeography, Palaeoclimatology, Palaeoecology* **127**: 1–20.

Brooks, R. A. (1991). "Intelligence Without Representation," *Artificial Intelligence* **47**: 139–59.

Burian, R. M. and Richardson, R. C. (1996). "Form and Order in Evolutionary Biology," in Boden, M. (Ed.), *The Philosophy of Artificial Life*. Oxford, Oxford University Press.

Buss, L. (1987). *The Evolution of Individuality*. Princeton, Princeton University Press.

Byrne, R. (1993). "The Meaning of 'Awareness,'" *New Ideas in Psychology*. **11**: 347–50.

Byrne, R. (1994). "The Evolution of Intelligence," in Slater, P. J. B. and Halliday, T. R. (Eds.), *Behaviour and Evolution*. Cambridge, Cambridge University Press.

Byrne, R. (1995a). "The Ape Legacy," in Goody, E. (Ed.), *Social Intelligence and Interaction*. Cambridge, Cambridge University Press.

Byrne, R. (1995b). *The Thinking Ape: Evolutionary Origins of Intelligence*. Oxford, Oxford University Press.

Byrne, R. and Whiten, A. (Eds.) (1988). *Machiavellian Intelligence: Social Expertise and the Evolution of Intellect in Monkeys, Apes and Humans*. Oxford, Oxford University Press.

Byrne, R. and Whiten, A. (1992). "Cognitive Evolution in Primates: Evidence from Tactical Deception," *Man* **27**: 609–25.

Cairns-Smith, G. (1982). *Genetic Takeover and the Mineral Origins of Life*. Cambridge, Cambridge University Press.

Camhi, J. M. (1984). *Neuroethology*. Sunderland, Sinauer.

Cartwright, N. (1979). "Causal Laws and Effective Strategies," *Nous* **13**: 419–37.

Cheney, D. and Seyfarth, R. (1990). *How Monkeys See the World*. Chicago, Chicago University Press.

Cheney, D. and Seyfarth, R. (1991). "Reading Minds or Reading Behaviour? Tests for a Theory of Mind in Monkeys," in Whiten, A. (Ed.), *Natural Theories of Mind: Evolution, Development and the Simulation of Everyday Mindreading*. Oxford, Blackwell.

Clark, A. (1996). "Happy Couplings: Emergence and Explanatory Interlock." In Boden, M. (Eds.), *The Philosophy of Artificial Life*. Oxford, Oxford University Press.

References

Clark, A. (1997). *Being There: Putting Brain, Body, and World Together Again.* Cambridge, Mass., MIT Press.

Clayton, D. and Harvey, P. (1993). "Hanging Nests on a Phylogenetic Tree," *Current Biology* **3**: 882–3.

Colwell, R. K. (1981). "Group Selection Implicated in the Evolution of Female Biased Sex Ratios," *Nature* **290**: 401–4.

Colwell, R. (1992). "Niche: A Bifurcation in the Conceptual Lineage of the Term," in Keller, E. F. and Lloyd, E. (Eds.), *Keywords in Evolutionary Biology.* Cambridge, Mass., Harvard University Press.

Connell, J. H. (1978). "Diversity in Tropical Rain Forests and Coral Reefs," *Science* **199**: 1302–10.

Cooper, G. (1993). "The Competition Controversy in Community Ecology," *Biology and Philosophy* **8**: 359–84.

Cooper, G. (1998). "Generalizations in Ecology: A Philosophical Taxonomy," *Biology and Philosophy* **13**: 555–86.

Cronin, H. (1991). *The Ant and the Peacock: Altruism and Sexual Selection from Darwin to Today.* Cambridge, Cambridge University Press.

Currie, G. (1995). "Visual Imagery as the Simulation of Vision," *Mind and Language* **10**: 25–44.

Currie, G. (1996). "Simulation-Theory, Theory-Theory and the Evidence from Autism," in Carruthers, P. and Smith, P. (Eds.), *Theories of Theories of Mind.* Cambridge, Cambridge University Press.

Damuth, J. (1985). "Selection Among Species: A Formulation in Terms of Natural Functional Units," *Evolution* **39**: 1132–46.

Damuth, J. and Heisler, L. (1988). "Alternative Formulations of Multilevel Selection," *Biology and Philosophy* **3**: 407–30.

Darwin, C. (1859/1964). *On the Origin of Species: A Facsimile of the First Edition.* Cambridge, Mass., Harvard University Press.

Dawkins, R. (1976/2nd ed. 1989). *The Selfish Gene.* Oxford, Oxford University Press.

Dawkins, R. (1982). *The Extended Phenotype.* Oxford, Oxford University Press.

Dawkins, R. (1986). *The Blind Watchmaker.* New York, W.W. Norton.

Dawkins, R. (1989). "The Evolution of Evolvability," in Langton, C. (Ed.), *Artificial Life VI Santa Fe Institute Studies in the Sciences of Complexity.* Redwood City, Calif., Addison-Wesley.

Dawkins, R. (1994). "Burying the Vehicle," *Behavioral and Brain Sciences* **17**: 617.

Dennett, D. C. (1983). "Intentional Systems in Cognitive Ethology: The 'Panglossian Paradigm' Defended," *Behavioural and Brain Sciences* **6**: 343–90.

Dennett, D. C. (1991a). *Consciousness Explained.* Boston, Little, Brown and Co.

Dennett, D. C. (1991b). "Real Patterns," *Journal of Philosophy* **88**: 27–51.

Dennett, D. C. (1994). "E Pluribus Unum?" *Behavioral and Brain Sciences* **17**: 617–18.

Dennett, D. C. (1995). *Darwin's Dangerous Idea.* New York, Simon and Shuster.

Depew, D. and Weber, B. H. (1995). *Darwinism Evolving: Systems Dynamics and the Genealogy of Natural Selection.* Cambridge, Mass., MIT Press.

Devitt, M. (1996). *Coming to Our Senses: A Naturalistic Program for Semantic Localism.* Cambridge, Cambridge University Press.

References

De Waal, F. (1982). *Chimpanzee Politics: Power and Sex Amongst the Apes*. New York, Harper and Row.

De Waal, F. (1989). *Peacemaking Among Primates*. Cambridge, Mass., Harvard University Press.

Diamond, J. M. (1975). "Assembly of Species Communities," in Cody, M. L. and Diamond, J. M. (Eds.), *Ecology and Evolution of Communities*. Cambridge, Mass., Harvard University Press.

Diamond, J. (1992). *The Third Chimpanzee*. New York, Harper/Collins.

Dickinson, A. (1985). "Actions and Habits: The Development of Behavioural Autonomy," *Philosophical Transactions of the Royal Society, London*, Series B **308**: 67–78.

Dickinson, A. and Balleine, B. (1993). "Actions and Responses: The Dual Psychology of Behaviour," in Eilan, N., McCarthy, R., and Brewer, M. W. (Eds.), *Problems in the Philosophy and Psychology of Spatial Representation*. Oxford, Blackwell.

Dickinson, A. and Balleine, B. W. (2000). "Causal Cognition and Goal Directed Action," in Heyes, C. and Huber, L. (Eds.), *The Evolution of Cognition*. Cambridge, Mass., MIT Press.

Dickinson, A. and Shanks, D. (1995). "Instrumental Action and Causal Representation," in Sperber, D., Premack, D., and Premack, A. J. (Eds.), *Causal Cognition: A Multidisciplinary Debate*. Oxford, Clarendon Press.

Dickison, M. (1992). "The Death of the Organism or The Selfish Nest." Paper to Australasian Association of Philosophy, New Zealand Division, Dunedin, July.

Dretske, F. (1972). "Contrastive Statements," *Philosophical Review* **81**: 411–37.

Dretske, F. (1981). *Knowledge and the Flow of Information*. Oxford, Blackwell.

Dretske, F. (1988). *Explaining Behavior: Reasons in a World of Causes*. Cambridge, Mass., MIT Press.

Dugatkin, L. A. and Reeve, H. K. (1994). "Behavioral Ecology and Levels of Selection: Dissolving the Group Selection Controversy," *Advances in the Study of Behavior* **23**: 101–33.

Dunbar, R. (1996). *Grooming, Gossip and the Evolution of Language*. London, Faber and Faber.

Dunbar, R. I. (1998). "The Social Brain Hypothesis," *Evolutionary Anthropology* **6**: 178–90.

Dupre, J. (1993). *The Disorder of Things*. Cambridge, Mass., Harvard University Press.

Eldredge, N. (1985a). *Time Frames*. New York, Simon & Schuster.

Eldredge, N. (1985b). *The Unfinished Synthesis*. Oxford, Oxford University Press.

Eldredge, N. (1989). *Macroevolutionary Dynamics*. New York, McGraw Hill.

Eldredge, N. (1995). *Reinventing Darwin*. New York, John Wiley and Son.

Evans, C., Evans, L., and Marler, P. (1993). "On the Meaning of Alarm Calls: Functional Reference in an Avian Vocal System," *Animal Behaviour* **46**: 23–38.

Evans, C., Macedonia, J. M., and Marler, P. (1993). "Effects of Apparent Size and Speed on the Response of Chickens, *Gallus gallus*, to Computer-generated Simulations of Aerial Predators," *Animal Behaviour* **46**: 1–11.

292

References

Evans, C. S. and Marler, P. (1995). "Language and Animal Communication: Parallels and Contrasts," in Roitblat, H. and Arcady-Meyer, J. (Eds.), *Comparative Approaches to Cognitive Science*. Cambridge, Mass., MIT Press.

Fodor, J. A. (1986). "Why Paramecia Don't Have Mental Representations," *Midwest Studies in Philosophy* **10**: 3–24.

Fodor, J. (1990). *A Theory of Content and Other Essays*. Cambridge, Mass., MIT Press.

Fogle, T. (1990). "Are Genes Units of Inheritance?" *Biology and Philosophy* **5**: 349–72.

Galef, B. (1992). "The Question of Animal Culture," *Human Nature* **3**: 157–78.

Garfinkel, A. (1981). *Forms of Explanation*. New Haven, Yale University Press.

Gaston, K. J. and Blackburn, T. M. (1999). "A Critique for Macroecology," *Oikos* **84**: 353–68.

Ghiselin, M. (1974). "A Radical Solution to the Species Problem," *Systematic Zoology* **23**: 536–44.

Gilinsky, N. L. (1986). "Species Selection as a Causal Process," *Evolutionary Biology* **20**: 249–73.

Glover, D., Gonzales, C., and Raff, J. (1993). "The Centrosome," *Scientific American* **268**: 32–9.

Godfrey-Smith, P. (1991a). "Signal, Decision, Action," *Journal of Philosophy* **88**: 709–22.

Godfrey-Smith, P. (1991b). *Teleonomy and the Philosophy of Mind*. Ph.D. Dissertation, University of California, San Diego.

Godfrey-Smith, P. (1996). *Complexity and the Function of Mind in Nature*. Cambridge, Cambridge University Press.

Godfrey-Smith, P. (1999). "Adaptation and the Power of Selection," *Biology and Philosophy* **14**: 181–94.

Goldman, A. (1993). "The Psychology of Folk Psychology," *Behavioral and Brain Sciences* **16**: 15–28.

Goldman, A. (1995). "In Defense of Simulation Theory," in Davies, M. and Stone, T. (Eds.), *Folk Psychology: The Theory of Mind Debate*. Oxford, Blackwell.

Goode, R. and Griffiths, P. (1995). "The Misuse of Sober's Selection for/Selection of Distinction," *Biology and Philosophy* **10**: 99–108.

Goodwin, B. (1989). "Unicellular Morphogenesis," in Stein, W. D. and Bonner, F. (Eds.), *Cell Shape*. Academic Press, New York.

Gopnik, A. and Wellman, H. (1995). "Why the Child's Theory of Mind Really Is a Theory," in Davies, M. and Stone, T. (Eds.), *Folk Psychology: The Theory of Mind Debate*. Oxford, Blackwell.

Gordon, R. (1995). "Folk Psychology as Simulation," in Davies, M. and Stone, T. (Eds.), *Folk Psychology: The Theory of Mind Debate*. Oxford, Blackwell.

Gould, S. J. (1980a). *The Panda's Thumb*. New York, W.W. Norton.

Gould, S. J. (1980b). "The Episodic Nature of Evolutionary Change," in Gould, S. J. (Ed.), *The Panda's Thumb*. New York, W.W. Norton.

Gould, S. J. (1980c). "The Return of the Hopeful Monster," in Gould, S. J. (Ed.), *The Panda's Thumb*. New York, W.W. Norton.

Gould, S. J. (1980d). "Is a New and General Theory of Evolution Emerging?" *Paleobiology* **6**: 119–30.

Gould, S. J. (1980e). "Caring Groups and Selfish Genes," in Gould, S. J. (Ed.), *The Panda's Thumb*. New York, W.W. Norton.

Gould, S. J. (1983). "The Meaning of Punctuated Equilibrium and Its Role in Validating a Hierarchical Approach to Macroevolution," *Scientia* **1**: 135–57.

Gould, S. J. (1985). "The Paradox of the First Tier: An Agenda for Paleobiology," *Paleobiology* **11**: 2–12.

Gould, S. J. (1989). *Wonderful Life: The Burgess Shale and the Nature of History*. New York, W.W. Norton.

Gould, S. J. (1991). "Of Kiwi Eggs and the Liberty Bell," in Gould, S. J. (Ed.), *Bully for Brontosaurus*. New York, W.W. Norton.

Gould, S. J. (1995). "A Task for Paleobiology at the Threshold of Majority," *Paleobiology* **21**: 1–14.

Gould, S. J. and Lewontin, R. (1978). "The Spandrels of San Marco and the Panglossian Paradigm: A Critique of the Adaptationist Program," *Proceedings of the Royal Society, London* **205**: 581–98.

Grafen, A. (1982). "How Not To Measure Inclusive Fitness," *Nature* **298**: 425–6.

Grant, T. (1989). *The Platypus: A Unique Mammal*. Kensington, Sydney, New South Wales University Press.

Gray, R. (1992). "The Death of the Gene," in Griffiths, P. (Ed.), *Trees of Life: Essays in the Philosophy of Biology*. Dordrecht, Kluwer.

Gray, R., Griffiths, P., and Oyama, S. (Eds.) (2000). *Cycles of Contingency*. Cambridge, Mass., MIT Press.

Griesemer, J. R. (1992). "Niche: Historical Perspectives," in Keller, E. F. and Lloyd, E. (Eds.), *Keywords in Evolutionary Biology*. Cambridge, Mass., Harvard University Press.

Griesemer, J. (forthcoming). "The Informational Gene and the Substantial Body: On the Generalization of Evolutionary Theory By Abstraction," in Cartwright, N. and Jones, M. (Eds.), *Varieties of Idealisation*. Amsterdam, Poznan Studies in the Philosophy of the Sciences and the Humanities.

Griffin, D. (1991). "Progress Towards a Cognitive Ethology," in Ristau, C. (Ed.), *Cognitive Ethology*. Hillsdale, LEA Press.

Griffiths, P. (1994). "Cladistic Classification and Functional Explanation," *Philosophy of Science* **61**: 206–27.

Griffiths, P. (1996). "The Historical Turn in the Study of Adaptation," *British Journal of the Philosophy of Science* **47**: 511–32.

Griffiths, P. and Gray, R. (1993, July). "Individuating Developmental Systems," *Paper to the International Society for the History, Philosophy and Social Studies of Biology*, Boston.

Griffiths, P. and Gray, R. (1994). "Developmental Systems and Evolutionary Explanation," *Journal of Philosophy* **91**: 277–304.

Hamilton, W. (1971). "The Genetical Evolution of Social Behavior," in Williams, G. C. (Ed.), *Group Selection*. Chicago, Aldine.

Hamilton, W. (1975). "Innate Social Aptitudes in Man: An Approach from Evolutionary Genetics," in Fox, R. (Ed.), *Biosocial Anthropology*. London, Malaby Press.

Hendriks-Jansen, H. (1996). "In Praise of Interactive Emergence; or Why Explanation Doesn't Have to Wait for Implementations," in Boden, M. (Ed.), *The Philosophy of Artificial Life*. Oxford, Oxford, University Press.

Heyes, C. M. (1993a). "Anecdotes, Training, Trapping and Triangulating: Do Animals Attribute Mental States?" *Animal Behaviour* **46**: 177–88.

Heyes, C. M. (1993b). "Imitation, Culture and Cognition," *Animal Behaviour* **46**: 999–1010.

Heyes, C. M. (1994a). "Reflections on Self-Recognition in Primates," *Animal Behaviour* **47**: 909–19.

Heyes, C. M. (1994b). "Social Cognition in Primates," in Mackintosh, N. J. (Ed.), *Handbook of Perception and Cognition*. Vol. 9. New York, Academic Press.

Heyes, C. M. (1998). "Theory of Mind in Non-Human Primates," *Behavioral and Brain Sciences* **21**: 101–48.

Heyes, C. M., and Dickinson, A. (1990). "The Intentionality of Animal Action," *Mind and Language* **5**: 87–104.

Hoffman, A. (1989). *Arguments on Evolution*. Oxford, Oxford University Press.

Holldobler, B. and Wilson, E. O. (1990). *The Ants*. Cambridge, Mass., Harvard University Press.

Horan, B. H. (1989). "Functional Explanations in Sociobiology," *Biology and Philosophy* **4**: 131–205.

Hull, D. (1978). "A Matter of Individuality," *Philosophy of Science* **45**: 335–60. [Reprinted in Sober, E. (1984). *Conceptual Issues in Evolutionary Biology*. Cambridge, MIT Press.]

Hull, D. (1981). "Units of Evolution: A Metaphysical Essay," in Jensen, R. and Harre, R. (Eds.), *The Philosophy of Evolution*. Brighton, Harvester.

Hull, D. (1987). "Genealogical Actors in Ecological Roles," *Biology and Philosophy* **2**: 168–84. [Reprinted in Hull, D. (1989). *The Metaphysics of Evolution*. Albany, State University of New York Press.]

Hull, D. (1988a). *Science as a Process*. Chicago, Chicago University Press.

Hull, D. (1988b). "Interactors and Vehicles," in Plotkin, H. (Ed.), *The Role of Behavior in Evolution*. Cambridge, Mass., MIT Press.

Hull, D. (1989). *The Metaphysics of Evolution*. Albany, State University of New York Press.

Hull, D. (1994). "Taking Vehicles Seriously," *Behavioral and Brain Sciences* **17**: 627–8.

Hutchinson, G. E. (1959). "Homage to Santa Rosalia; or, Why Are There So Many Kinds of Animals?" *American Naturalist* **93**: 145–59.

Hutchinson, G. E. (1965). *The Ecological Theater and the Evolutionary Play*. New Haven, Yale University Press.

Hutchinson, G. E. (1978). *Introduction to Population Ecology*. New Haven, Yale University Press.

Ivany, L. (1996). "Coordinated Stasis or Coordinated Turnover? Exploring Intrinsic vs. Extrinsic Controls on Pattern," *Palaeo: Palaeogeography, Palaeoclimatology, Palaeoecology* **127**: 239–56.

Jackson, F. and Pettit, P. (1992). "In Defence of Explanatory Ecumenicalism," *Economics and Philosophy* **8**: 1–21.

References

James, F., Johnston, R. F., et al. (1984). "The Grinellian Niche of the Wood Thrush," *American Naturalist* **124**: 17–47.

Jarrold, C., Carruthers, P., et al. (1994). "Pretend Play: Is It Metarepresentational?" *Mind and Language* **9**(4): 445–68.

Jax, K., Jones, C. G., et al. (1998). "The Self-Identity of Ecological Units," *Oikos* **82**: 253–64.

Johnston, T. (1987). "The Persistence of Dichotomies in the Study of Behavioural Development," *Developmental Review* **7**: 149–82.

Johnston, T. (1988). "Developmental Explanation and The Ontogeny of Birdsong: Nature/Nuture Redux," *Behavioral and Brain Sciences* **11**: 617–63.

Jolly, A. (1988). "The Evolution of Purpose," in Byrne, R. and Whiten, A. (Eds.), *Machiavellian Intelligence: Social Expertise and the Evolution of Intellect in Monkeys, Apes and Humans.* Oxford, Oxford University Press.

Jolly, A. (1991). "Conscious Chimpanzees? A Review of Recent Literature," in Ristau, C. (Ed.), *Cognitive Ethology.* Hillsdale, LEA Press.

Jones, C., Lawton, J., et al. (1997). "Positive and Negative Effects of Organisms as Physical Ecosystems Engineers," *Ecology* **78**: 1946–57.

Keller, L. and Ross, K. G. (1993). "Phenotypic Plasticity and 'Cultural Transmission' in the Fire Ant *Solenopsis invicta*," *Behavioural Ecology and Sociobiology* **33**: 121–9.

Kettlewell, H. B. D. (1973). *The Evolution of Melanism.* New York, Oxford University Press.

Kingsland, S. (1985). *Modeling Nature: Episodes in the History of Population Ecology.* Chicago, Chicago University Press.

Kitcher, P. (1985). *Vaulting Ambition: Sociobiology and the Quest for Human Nature.* Cambridge, Mass., MIT Press.

Kitcher, P. (1987). "Why Not the Best?" in Dupre, J. (Ed.), *The Latest on The Best.* Cambridge, Mass., MIT Press.

Kitcher, P. (1989). "Some Puzzles about Species," in Ruse, M. (Ed.), *What the Philosophy of Biology Is.* Dordrecht, Kluwer.

Krebs, J. and Dawkins, R. (1984). "Animal Signals, Mind-Reading and Manipulation," in Krebs, J. R. and Davies, N. B. (Eds.), *Behavioural Ecology: An Evolutionary Approach.* Oxford, Blackwell Scientific.

Lack, D. (1966). *Population Studies of Birds.* Oxford, Oxford University Press.

Laland, K. N., Odling-Smee, J., et al. (forthcoming). "Niche Construction, Biological Evolution and Cultural Change," *Behavioral and Brain Sciences.*

Lawton, J. H. (1999). "Are There General Laws in Ecology?" *Oikos* **84**: 177–92.

Leslie, A. M. (1994). "Pretending and Believing: Issues in the Theory of ToMM," *Cognition* **50**: 211–38.

Leslie, A. and German, T. (1995). "Knowledge and Ability in 'Theory of Mind': One-eyed Overview of Debate," in Davies, M. and Stone, T. (Eds.), *Mental Simulation: Evaluations and Applications.* Oxford, Blackwell.

Levins, R. and Lewontin, R. (1985). *The Dialectical Biologist.* Cambridge, Mass., Harvard University Press.

References

Lewontin, R. C. (1982). "Organism and Environment," in Plotkin, H. C. (Ed.), *Learning, Development, and Culture.* New York, Wiley.

Lewontin, R. C. (1983). "The Organism as the Subject and Object of Evolution," *Scientia* **118**: 65–82.

Lewontin, R. (1985). "The Organism as the Subject and Object of Evolution," in Levins, R. and Lewontin, R. (Eds.), *The Dialectical Biologist.* Cambridge, Mass., Harvard University Press.

Linsenmair, K. E. (1987). "Kin Recognition in Subsocial Arthropods, in Particular, the Desert Isopod *Hemilepistus reaumuri*," in Fletcher, D. J. and Michener, C. D. (Eds.), *Kin Recognition in Animals.* Chichester, John Wiley and Sons.

Lloyd, E. (1988). *The Structure and Confirmation of Evolutionary Theory.* Westport, Conn., Greenwood Press.

Lloyd, E. (1993). "Unit of Selection," in Keller, E. and Lloyd, E. (Eds.), *Keywords in Evolutionary Biology.* Cambridge, Mass., Harvard University Press.

Lloyd, J. E. (1984). "On Deception, a Way of All Flesh, and Firefly Signalling and Systematics," in Dawkins, R. and Ridley, M. (Eds.), *Oxford Surveys in Evolutionary Biology.* Vol. I. Oxford, Oxford University Press.

MacArthur, R. H. (1972). *Geographical Ecology: Patterns in the Distribution of Species.* New York, Harper and Row.

Majerus, M. and Hurst, G. (1993). "Weird Genetics? Evolution and Nonmendelian Genes," *Trends in Ecology and Evolution* **8**: 310–11.

Maynard Smith, J. (1964). "Group Selection and Kin Selection," *Nature* **201**: 1145–7.

Maynard Smith, J. (1982). *Evolution and the Theory of Games.* Cambridge, Cambridge University Press.

Maynard Smith, J. (1987). "How to Model Evolution," in Dupre, J. (Ed.), *The Latest on the Best.* Cambridge, Mass., MIT Press.

Maynard Smith, J. (1989). *Did Darwin Get It Right?* London, Chapman and Hall.

Maynard Smith, J. (forthcoming). "The Concept of Information in Biology," *Philosophy of Science* **67**.

Maynard Smith, J. and Szathmary, E. (1999). *The Origins of Life: From the Birth of Life to the Origins of Language.* Oxford, Oxford University Press.

Mayr, E. (1963). *Animal Species and Evolution.* Cambridge, Mass., Harvard University Press.

Mayr, E. (1976). *Evolution and The Diversity of Life.* Cambridge, Mass., Harvard University Press.

Mayr, E. (1988a). "Speciational Evolution through Punctuated Equilibria," in Mayr, E. *Towards a New Philosophy of Biology.* Cambridge, Mass., Harvard University Press.

Mayr, E. (1988b). *Towards a New Philosophy of Biology.* Cambridge, Mass., Harvard University Press.

Mayr, E. (1991). *One Long Argument: Charles Darwin and the Genesis of Modern Evolutionary Thought.* London, Penguin.

McFarland, D. J. (1996). "Animals as Cost-Based Robots," in Boden, M. (Ed.), *The Philosophy of Artificial Life.* Oxford, Oxford University Press.

McIntosh, R. (1992). "Competition: Historical Perspectives," in Keller, E. F. and

References

Lloyd, E. (Eds.), *Keywords in Evolutionary Biology*. Cambridge, Mass., Harvard University Press.

McNamara, K. (1987). "Australian Ammonites," *Australian Natural History* **22**(7): 332–6.

Meltzoff, A. and Gopnik, A. (1993). "The Role of Imitation in Understanding Persons and Developing a Theory of Mind," in Baron-Cohen, S., Tager-Flusberg, H. and Cohen, D. J. (Eds.), *Understanding Other Minds: Perspectives from Autism*. Oxford, Oxford University Press.

Michod, R. (1984). "The Theory of Kin Selection," in Brandon, R. and Burian, R. (Eds.), *Genes, Organisms, and Populations*. Cambridge, Mass., MIT Press.

Millikan, R. (1984). *Language, Thought and Other Biological Categories*. Cambridge, Mass., MIT Press.

Millikan, R. (1989a). "Biosemantics," *Journal of Philosophy* **86**: 281–97.

Millikan, R. (1989b). "In Defense of Proper Functions," *Philosophy of Science* **56**: 288–302.

Millikan, R. (1990). "Truth Rules, Hoverflies and the Kripke-Wittgenstein Paradox," *Philosophical Review* **94**: 323–53.

Millikan, R. (1991). "Speaking Up for Darwin," in Loewer, B. and Rey, R. (Eds.), *Meaning in Mind: Fodor and his Critics*. Oxford, Blackwell.

Morgan, N. and Baumann, P. (1994). "Phylogenetics of Cytoplasmically Inherited Microorganisms of Arthropods," *Trends in Ecology and Evolution* **9**: 15–20.

Morris, P., Ivany, L., et al. (1995). "The Challenge of Paleoecological Stasis: Reassessing Sources of Evolutionary Stability," *Proceedings of the National Academy of Science, USA* **92**: 11269–73.

Moss, L. (1992). "A Kernel of Truth? On the Reality of the Genetic Program," in Hull, D., Forbes, M., and Okruhlik, K. (Eds.), *Proceedings of the Philosophy of Science Association*. Vol. I. East Lansing, Mich., Philosophy of Science Association.

Nunney, L. (1989). "The Maintenance of Sex by Group Selection," *Evolution* **43**: 245–57.

Odling-Smee, F. J., Laland, K. N., et al. (1996). "Niche Construction," *American Naturalist* **147**: 641–8.

O'Hara, R. J. (1993). "Systematic Generalization, Historical Fate and the Species Problem," *Systematic Biology* **42**: 231–46.

Orzack, S. and Sober, E. (1994a). "How (Not) To Test an Optimality Model," *Trends In Ecology and Evolution* **9**: 265–7.

Orzack, S. and Sober, E. (1994b). "Optimality Models and the Test of Adaptationism," *American Naturalist* **143**: 361–80.

Otte, D. and Endler, J. (Eds.) (1989). *Speciation and Its Consequences*. Sunderland, Sinauer.

Oyama, S. (1985). *The Ontogeny of Information*. Cambridge, Mass., MIT Press.

Pinker, S. (1994). *The Language Instinct: How the Mind Creates Language*. New York, William Morrow and Co.

Povinelli, D. (1993). "Reconstructing the Evolution of Mind," *American Psychologist* **48**: 493–509.

298

References

Povinelli, D. J. and Cant, J. G. (1995). "Arboreal Clambering and the Evolution of Self-Conception," *Quarterly Review of Biology* **70**: 393–421.

Povinelli, D., Nelson, K., and Boysen, S. (1990). "Inferences about Guessing and Knowing by Chimpanzees," *Journal of Comparative Psychology* **104**: 203–10.

Premack, D. (1988). "'Does The Chimpanzee Have A Theory of Mind' Revisited," in Byrne, R. and Whiten, A. (Eds.), *Machiavellian Intelligence: Social Expertise and the Evolution of Intellect in Monkeys, Apes and Humans.* Oxford, Oxford University Press.

Premack, D. and Woodruff, G. (1978). "Does the Chimpanzee Have a Theory of Mind?" *Behavioral and Brain Sciences* **4**: 515–629.

Quine, W. V. O. (1953, 2nd ed. 1961). *From a Logical Point of View.* New York, Harper Torchbooks.

Raup, D. M. (1991). *Extinction: Bad Genes or Bad Luck?* New York, W.W. Norton.

Reilly, P. (1991). *The Lyrebird.* Kensington, Sydney, New South Wales University Press.

Ricklefs, R. E. (1989). "Speciation and Diversity: The Integration of Local and Regional Processes," in Otte, D. and Endler, J. A. (Eds.), *Speciation and Its Consequences.* Sunderland, Sinauer.

Ridley, M. (1986). *Evolution and Classification.* Longmans, London.

Ridley, M. (1989). "The Cladistic Solution to the Species Concept," *Biology and Philosophy* **4**: 1–16.

Ristau, C. (Ed.) (1991a). *Cognitive Ethology.* Hillsdale, LEA Press.

Ristau, C. (1991b). "Aspects of Cognitive Ethology of an Injury-Feigning Bird, the Piping Plover," in Ristau, C. (Ed.), *Cognitive Ethology.* Hillsdale, LEA Press.

Rosenberg, A. (1994). *Instrumental Biology or the Disunity of Science.* Chicago, Chicago University Press.

Roughgarden, J. and Pacala, S. (1989). "Taxon Cycle Among Anolis Lizard Populations: Review of Evidence," in Otte, D. and Endler, J. A. (Eds.), *Speciation and Its Consequences.* Sunderland, Sinauer.

Salmon, W. (Ed.) (1971). *Statistical Explanation and Statistical Relevance.* Pittsburgh, Pittsburgh University Press.

Salmon, W. (1984). *Scientific Explanation and the Causal Structure of the World.* Princeton, Princeton University Press.

Salthe, S. N. (1985). *Evolving Hierarchical Systems.* New York, Columbia University Press.

Salthe, S. and Eldredge, N. (1984). "Hierarchy and Evolution," *Oxford Surveys of Evolutionary Biology* **1**: 184–208.

Schoener, T. W. (1986). "Overview: Kinds of Ecological Communities – Ecology Becomes Pluralistic," in Diamond, J. and Case, T. W. (Eds.), *Community Ecology.* New York, Harper and Row.

Schoener, T. W. (1989). "The Ecological Niche," in Cherrett, J. M. (Ed.), *Ecological Concepts.* 29th Symposium of the British Ecological Society. Oxford, Blackwell Scientific Publications.

Schull, J. (1990). "Are Species Intelligent?" *Behavioral and Brain Sciences* **13**: 63–108.

Smith, K. (1992). "The New Problem of Genetics: A Response to Gifford," *Biology and Philosophy* **7**: 331–48.

Smith, K. (1993a). "Neo-Rationalism Versus Neo-Darwinism: Integrating Development and Evolution," *Biology and Philosophy* **7**: 431–52.

Smith, K. (1993b). "The Effects of Temperature and Daylength on the Rosa Polyphenism in the Buckeye Butterfly, *Precis coenia* (Lepidoptera: Nymphalidae)," *Journal of Research on the Lepidoptera* **30**: 225–36.

Smith, K. (1994). *The Emperor's New Genes: The Role of the Genome in Development and Evolution*. Ph.D. Dissertation, Duke University, Durham, N.C., USA.

Sober, E. (1984). *The Nature of Selection*. Cambridge, Mass., MIT Press.

Sober, E. (Ed.) (1984; 2nd ed. 1994). *Conceptual Issues in Evolutionary Biology*. Cambridge, Mass., MIT Press.

Sober, E. (1987). "What is Adaptationism?" in Dupre, J. (Ed.), *The Latest on the Best*. Cambridge, Mass., MIT Press.

Sober, E. (1990). "The Poverty of Pluralism: A Reply to Sterelny and Kitcher," *Journal of Philosophy* **87**: 151–8.

Sober, E. (1993). *Philosophy of Biology*. Boulder, Col., Westview Press.

Sober, E. and Lewinton, R. (1982). "Artifact, Cause and Genic Selection," *Philosophy of Science* **49**: 157–80.

Sober, E. and Wilson, D. S. (1998). *Unto Others: The Evolution of Altruism*. Cambridge, Mass., Harvard University Press.

Spier, E. and McFarland, D. (1998). "Learning to Do Without Cognition," in Pfeifer, R., Blumberg, B., Meyer, J.-A., and Wilson, S. (Eds.), *From Animals to Animats 5*. Cambridge, Mass., MIT Press.

Stanley, S. (1981). *The New Evolutionary Timetable*. New York, Basic Books.

Sterelny, K. (1990a). *The Representational Theory of Mind: An Introduction*. Oxford, Blackwell.

Sterelny, K. (1990b). "Animals and Individualism," in Hanson, P. (Ed.), *Information, Language and Cognition*: Vancouver Studies in Cognitive Science 1. Vancouver, UBC Press.

Sterelny, K. (1991). "Review of *Wonderful Life: The Burgess State and the Nature of History*," *Australasian Journal of Philosophy* **69**: 342–6.

Sterelny, K. (1995). "Basic Minds," *Philosophical Perspectives* **9**: 251–70.

Sterelny, K. (1997). "Navigating the Social World: Simulation versus Theory," *Philosophical Books* **37**: 11–29.

Sterelny, K. (1999a). "Supply-side Biology: A Critical Notice of Rudy Raff's *The Shape of Life*," *Metascience* **8**: 405–19.

Sterelny, K. (1999b). "Species as Ecological Mosaics," in Wilson, R. A. (Ed.), *Species: New Interdisciplinary Essays*. Cambridge, Mass., MIT Press.

Sterelny, K. (2000a). "Primate Worlds," in Heyes, C. and Huber, L. (Eds.), *Evolution of Cognition*. Cambridge, Mass., MIT Press.

Sterelny, K. (2000b). "Niche Construction, Developmental Systems and the Extended Replicator," in Gray, R., Griffiths, P., and Oyama, S. (Eds.), *Cycles of Contingency*. Cambridge, Mass., MIT Press.

Sterelny, K. (forthcoming). "Development, Evolution and Adaptation," *Philosophy of Science* (supplementary volume).

References

Sterelny, K. and Griffiths, P. (1999). *Sex and Death: An Introduction to Philosophy of Biology*. Chicago, University of Chicago Press.

Sultan, S. (1987). "Evolutionary Implications of Phenotypic Plasticity in Plants," in Hecht, M., Watson, B., and Prance, G. (Eds.), *Evolutionary Biology* 21. New York, Plenum Press.

Thompson, J. N. (1994). *The Coevolutionary Process*. Chicago, University of Chicago Press.

Thompson, J. N. (1999). "The Raw Material for Coevolution," *Oikos* **84**: 5–16.

Thornton, I. (1996). *Krakatau: The Destruction and Reassembly of an Island Ecosystem*. Cambridge, Mass., Harvard University Press.

Tinbergen, N. (1960). *The Herring Gull's World*. New York, Basic Books.

Tomasello, M. (2000). "Two Hypotheses about Primate Cognition," in Heyes, C. and Huber, L. (Eds.), *Evolution of Cognition*. Cambridge, Mass., MIT Press.

Turlings, T., Tomlinson, J., and Lewis, W. (1990). "Exploitation of Herbivore-Induced Plant Odors by Host-Seeking Parasitic Wasps," *Science* **250**: 1251–3.

Valentine, J. W. (1990). "The Macroevolution of Clade Shape," in Ross, R. and Allmon, W. (Eds.), *The Causes of Evolution: A Paleontological Perspective*. Chicago, Chicago University Press.

van Valen, L. (1973). "A New Evolutionary Law," *Evolutionary Theory* **1**: 1–30.

Vrba, E. S. (1984a). "Patterns in the Fossil Record and Evolutionary Processes," in Ho, M.-W. and Saunders, P. (Eds.), *Beyond Neo-Darwinism: An Introduction to the New Evolutionary Paradigm*. London, Academic Press.

Vrba, E. S. (1984b). "Evolutionary Pattern and Process in the Sister-Group *Alcelaphini-Aepycerotini* (Mammalia: Bovidae)," in Eldredge, N. and Stanley, S. (Eds.), *Living Fossils*. New York, Springer Verlag.

Vrba, E. (1984c). "What is Species Selection?" *Systematic Zoology* **33**: 318–28.

Vrba, E. (1993). "Turnover-Pulses, The Red Queen and Related Topics," *American Journal of Science* **293-A**: 418–52.

Vrba, E. S. (1995). "Species as Habitat-Specific Complex Systems," in Lambert, D. M. and Spencer, H. G. (Eds.), *Speciation and the Recognition Concept: Theory and Applications*. Baltimore, Johns Hopkins University Press.

Vrba, E. and Eldredge, N. (1984). "Individuals, Hierarchies and Processes: Towards a More Complete Evolutionary Theory," *Paleobiology* **10**: 146–71.

Wade, M. J. (1978). "A Critical Review of the Models of Group Selection," *Quarterly Review of Biology* **53**: 101–14.

Wade, M. J. (1985). "Soft Selection, Hard Selection, Kin Selection and Group Selection," *American Naturalist* **125**: 61–73.

Wagner, G. P. (1988). "The Vexing Role of Replicators in Evolution," *Biology and Philosophy* **3**: 322–36.

Waters, K. (1994). "Tempered Realism about the Forces of Selection," *Philosophy of Science* **58**: 553–73.

Weber, M. (1999). "The Aim and Structure of Ecological Theory," *Philosophy of Science* **66**: 71–93.

Weiner, J. (1994). *The Beak of the Finch: Evolution in Real Time*. London, Vintage.

References

Werren, J. H. (1991). "The Paternal Sex Ratio Chromosome of *Nasonia*," *American Naturalist* **137**: 392–402.

Whiten, A. (Ed.) (1991). *Natural Theories of Mind: Evolution, Development and the Simulation of Everyday Mindreading*. Oxford, Blackwell.

Whiten, A. (1996). "When Does Smart Behaviour-Reading Become Mind-Reading?" in Carruthers, P. and Smith, P. (Eds.), *Theories of Theories of Mind*. Cambridge, Cambridge University Press.

Whiten, A. and Byrne, R. (1988). "Tactical Deception in Primates," *Behavioral and Brain Sciences* **11**: 233–73.

Whiten, A. and Ham, R. (1992). "On the Nature and Evolution of Imitation in the Animal Kingdom," *Advances in the Study of Behavior* **21**: 239–83.

Whittaker, R. H. (1975). *Communities and Ecosystems*. New York, MacMillan.

Williams, G. C. (1966). *Adaptation and Natural Selection*. Princeton, Princeton University Press.

Williams, G. C. (1992). *Natural Selection: Domains, Levels and Challenges*. Oxford, Oxford University Press.

Wills, C. (1991). *The Wisdom of the Genes*. Oxford, Oxford University Press.

Wilson, D. S. (1980). *The Natural Selection of Populations and Communities*. Menlo Park, California, Benjamin/Cummings.

Wilson, D. S. (1983). "The Group Selection Controversy: History and Current Status," *Annual Review of Ecology and Systematics* **14**: 159–87.

Wilson, D. S. (1989). "Levels of Selection: An Alternative to Individualism in Biology and the Social Sciences," *Social Networks* **11**: 257–72.

Wilson, D. S. (1990). "Weak Altruism, Strong Group Selection," *Oikos* **59**: 135–40.

Wilson, D. S. (1992). "Complex Interactions in Metacommunities, with Implications for Biodiversity and Higher Levels of Selection," *Ecology* **73**: 1984–2000.

Wilson, D. S. (1993). "Group Selection," in Keller, E. and Lloyd, E. (Eds.), *Keywords in Evolutionary Biology*. Cambridge, Mass., Harvard University Press.

Wilson, D. (1999). "A Critique of R. D. Alexander's Views of Group Selection," *Biology and Philosophy* **14**: 431–49.

Wilson, D. S. (1997). "Incorporating Group Selection into the Adaptationist Program: A Case Study Involving Human Decision Making," in Simpson, J. and Kendrick, D. (Eds.), *Evolutionary Approaches in Personality and Social Psychology*. Hillsdale, N.J., Erlbaum Press.

Wilson, D. S. and Sober, E. (1989). "Reviving the Superorganism," *Journal of Theoretical Biology* **136**: 332–56.

Wilson, D. S. and Sober, E. (1994). "Reintroducing Group Selection to Human Behavioral Sciences," *Behavioral and Brain Sciences* **17**: 585–654.

Wimsatt, W. (1981). "The Units of Selection and the Structure of the Multi-Level Genome," in Asquith, P. and Giere, R. (Eds.), *Proceedings of the PSA 1980, Volume 2*. East Lansing, Mich., Philosophy of Science Association.

Wynne-Edwards, V. C. (1962). *Animal Dispersion in Relation to Social Behaviour*. Edinburgh, Liver and Boyd.

Wynne-Edwards, V. C. (1986). *Evolution through Group Selection*. Oxford, Blackwell Scientific.

302

References

Zahavi, A. and Zahavi, A. (1997). *The Handicap Principle: A Missing Piece of Darwin's Puzzle.* Oxford, Oxford University Press.

Zobel, M. (1997). "The Relative Role of Species Pools in Determining Plant Species Richness: An Alternative Explanation of Species Coexistence?" *Trends in Ecology and Evolution* **12**: 266–9.

References

Zhou, Weirei, Wong, Peter X. (1996) *Chinese approach to learning* (first edition). Pearson Prentice Hall.

Zhao, Ar (1998) "*Pedagogy, Knowledge...*". Pedagogium Press.

Daniel, Taylor, Xu. *Arithmetic Explanation of Speed Assessment,* languages and exposition, pp 2-26.

Index

Index